Holography and Its Applications

Edited by Michael R. Wang

Published in London, United Kingdom

Holography and Its Applications
http://dx.doi.org/10.5772/intechopen.1003388
Edited by Michael R. Wang

Contributors
A. Qadoos, Abdul Waheed Anwar, Alfonso López-Martínez, Armando Santos-Amador, Camilo Fernandez, Diana I. Ortiz-Esquivel, Diego Palacios, Hui Gao, Humberto González-Hernández, Inmaculada Pascual, Junaid Hassan, Ma. A. Araiza-Esquivel, Marta Morales-Vidal, Mehrunisa Moin, Michael R. Wang, Misael Torres-García, Muhammad Moin, Raúl E. Balderas-Navarro, Tomás Lloret

First published in London, United Kingdom, 2025 by IntechOpen
IntechOpen is the global imprint of INTECHOPEN LIMITED, registered in England and Wales, registration number: 11086078, 167-169 Great Portland Street, London, W1W 5PF, United Kingdom

For EU product safety concerns: IN TECH d.o.o., Prolaz Marije Krucifikse Kozulić 3, 51000 Rijeka, Croatia, info@intechopen.com or visit our website at intechopen.com.

British Library Cataloguing-in-Publication Data
A catalogue record for this book is available from the British Library

Holography and Its Applications
Edited by Michael R. Wang
p. cm.
Print ISBN 978-0-85014-853-4
Online ISBN 978-0-85014-852-7
eBook (PDF) ISBN 978-0-85014-854-1

Meet the editor

Michael R. Wang is a Professor of the Department of Electrical and Computer Engineering, University of Miami. He received his Ph.D. degree in 1992 from the Department of Electrical Engineering, University of California, Irvine. His research is focused on integrated photonic devices, optical interconnects, holography, lithography, spectral imaging, and optical coherence tomography. Dr. Wang has been a principal investigator and/or project leader in more than 50 US government-sponsored projects. He has authored/co-authored more than 210 journal papers, proceedings, and conference presentations. He is a fellow of SPIE, a senior member of OPTICA, and a member of ARVO.

Contents

Preface

Holography is a technique that utilizes the coherence property of light to record the 3D images of objects. The interference of the object and reference beams from a common light source forms the interference pattern. In traditional analog holography, the intensity and phase information of the interference pattern is stored in a volume light-sensitive holographic medium, which can be a crystal or a film. The readout using a reference light beam can recover the 3D image of the object, namely the hologram. Significant research and development have been focused on holographic recording techniques and recording media that can improve the hologram recording speed, image quality, minimization of hologram readout erasure degradation, and hologram storage lifetime.

Digital holography is an advanced technique that allows the recording, processing, and reconstruction of 3D images of objects using digital sensors such as a digital camera. While some intensity and phase information may be lost in a single exposure recording in the 2D camera pixels, multi-frame recording with tuning of camera positions and/or reference light phases can help regain the lost information for the hologram readout using advanced image processing and phase retrieval computations. Digital holography is attractive in user-friendly 3D rotational image display, long storage lifetime, and online image delivery.

The majority of holographic applications have been found in medical imaging, security and anti-counterfeiting, three-dimensional imaging and displays, industrial non-destructive testing and surface profiling, virtual reality and augmented reality, holographic teleconferencing, and product design and prototyping. This book presents a few recent advancements in holography and its applications.

Michael R. Wang, PhD
Department of Electrical and Computer Engineering,
University of Miami,
Florida, USA

Chapter 1

Revolutionizing Tomorrow: Embracing the Cutting-Edge of Holographic Technology

Camilo Fernandez

Abstract

This chapter delves into the transformative realm of holographic technology, examining its evolution, current state, and potential impact on various industries and everyday life. It provides a comprehensive overview of holography, tracing its origins from scientific experimentation to its widespread applications in entertainment, communication, healthcare, and more. The chapter explores the underlying principles of holographic displays, detailing advancements in hardware and software that have propelled this technology forward. Additionally, it analyzes the implications of holographic technology on user experience, immersive storytelling, data visualization, and beyond. By exploring the multifaceted dimensions of holography, this chapter aims to provide insight into its promising future and potential societal implications.

Keywords: holography, technological innovation, futuristic technology, holographic displays, emerging technologies

1. Introduction

In recent years, holographic technology has emerged as a revolutionary force, poised to reshape various sectors and redefine everyday experiences. This article provides an in-depth exploration of holography, charting its journey from initial scientific experiments to its current, expansive applications. From entertainment and communication to healthcare and beyond, holography has infiltrated numerous domains, offering novel solutions and enhancing user experiences.

This discussion will cover the foundational principles of holographic displays, highlighting significant advancements in both hardware and software that have driven this technology forward. Companies are now developing astonishing 3D visual holograms, pushing the boundaries of what is possible and setting new standards for visual realism and interactivity.

Moreover, we will examine the broader implications of holography, particularly in terms of immersive storytelling, data visualization, and user engagement. By delving into the multifaceted aspects of holographic technology, this article aims to illuminate its transformative potential and forecast its impact on society.

Through this comprehensive analysis, we hope to provide readers with a clearer understanding of the promising future that holography holds.

2. Holographic technology: Shaping the future of visual communication

2.1 The evolution of holography: From concept to reality

Holographic technology has a rich history, beginning as a theoretical concept and evolving through decades of scientific research and experimentation. The journey of holography from its inception to its current state is a testament to human ingenuity and technological advancement. Initially, holography was a niche area within the field of optics, primarily confined to laboratories and specialized research institutions.

The first practical demonstration of holography came in the mid-twentieth century with the invention of the laser, which provided the coherent light source necessary for creating holograms. Early holograms were static and required meticulous production processes, limiting their practical applications. However, these initial efforts laid the groundwork for future developments, proving that three-dimensional imaging was possible.

2.2 Humble origins of holography

The concept of holography dates back to the early twentieth century and is rooted in the work of Hungarian-British physicist Dennis Gabor [1]. In 1947, Gabor developed the theory of holography while working to improve the resolution of electron microscopes. His groundbreaking idea was to capture all the information about a light field, including phase and amplitude, which would later enable the reconstruction of a three-dimensional image. For his pioneering work, Gabor was awarded the Nobel Prize in Physics in 1971.

However, the practical realization of holography required a coherent light source, which did not become available until the invention of the laser in 1960 by Theodore Maiman [2]. The advent of laser technology provided the necessary tools to produce clear and detailed holograms, setting the stage for further advancements in the field.

2.3 Early developments and applications

The 1960s and 1970s saw significant progress in holography, with researchers creating the first practical holograms [3]. Early applications were primarily scientific and industrial, including non-destructive testing of materials, data storage, and security features on banknotes and credit cards. These early holograms were static and required meticulous production processes, which limited their widespread use.

One notable milestone was the development of the transmission hologram, which records and reconstructs light passing through a holographic plate. This type of hologram laid the foundation for future innovations and demonstrated the potential of holography to capture and display three-dimensional images.

3. Advancements in holographic displays

Recent advancements in holographic technology have been nothing short of remarkable. Modern holographic displays have moved beyond the constraints of their early predecessors, leveraging sophisticated hardware and software to create dynamic, interactive, and highly realistic 3D images [4]. Innovations such as digital holography, spatial light modulators, and advanced projection systems have significantly enhanced the quality and accessibility of holographic displays.

Companies at the forefront of this technology are developing astonishing 3D visual holograms that offer unparalleled depth and clarity. These advancements are not only enhancing user experiences but also opening new avenues for applications across various industries. From virtual meetings and immersive training sessions to cutting-edge medical imaging, the potential uses of holography are expanding rapidly.

3.1 Impact on various industries

The impact of holographic technology is being felt across a multitude of sectors. In the entertainment [5] industry, holography is revolutionizing the way audiences experience concerts, movies, and video games, offering immersive experiences that were previously unimaginable. In communication [6], holographic displays are enabling more engaging and lifelike interactions, potentially transforming virtual meetings and teleconferences.

Healthcare [7] is another field where holography is making significant strides. Advanced holographic imaging techniques are providing doctors with detailed, three-dimensional views of patient anatomy, enhancing diagnostic accuracy and improving surgical planning. Furthermore, holography is being used in educational settings to create interactive learning environments, helping students grasp complex concepts through visual and experiential learning.

In addition, Holograms improve user privacy and security measures by enabling touchless interaction, preventing physical traces of passwords, and utilizing dynamic layouts and motion tracking to secure data input against unauthorized access.

4. Advances in holographic display technology: The leading edge with HoloConnects, Aska 3d, and Light Field Lab

4.1 HoloConnects

At the forefront of holographic technology, HoloConnects [8] has developed the Holobox, an Artificial intelligence (AI)-powered holographic display system. The Holobox captures a subject's image against a white backdrop and projects it into a specially designed box, creating a customizable three-dimensional hologram. This system is enhanced with AI-powered graphics and atmospheric effects, making the hologram experience more lifelike and interactive. HoloConnects [8] is deploying its technology across various industries, including hospitality, healthcare [7], and retail. For instance, in the hospitality industry, Holobox units are being installed in hotels to serve as interactive check-in kiosks. In healthcare [7], Holobox is connecting doctors with patients in remote areas, allowing for real-time medical consultations through live holograms, HoloConnects [8] has developed a sophisticated process to create high-quality, AI-powered holographic displays through their proprietary system, the Holobox. Here is a detailed breakdown of how they develop these holograms.

4.1.1 Data acquisition and preparation

HoloConnects [8] begins by collecting high-resolution, multi-angle images or video footage of the subject or object that will be turned into a hologram. This data can be captured using specialized cameras and sensors that record the subject from multiple perspectives.

4.1.2 AI-powered processing

The captured data is then processed using advanced AI algorithms. HoloConnects [8] employs machine-learning techniques to analyze and synthesize the visual information. This step is crucial for ensuring that the hologram accurately represents the original subject, with lifelike details and proper depth perception.

To create high-resolution holograms, *HoloConnects* utilizes specialized cameras and sensors designed to capture multi-angle images or video footage of a subject or object. These cameras are equipped with the following technical capabilities:

1. *High-resolution imaging*: The cameras used in the process have the capability to capture ultra-high-definition images, often in 4 K or higher resolution, to ensure that the holographic reproduction has intricate details and clarity.

2. *Multi-angle capture*: Multiple cameras are arranged around the subject at various positions to record from different perspectives simultaneously. This setup allows for a full 360-degree capture of the subject, which is critical for generating a three-dimensional hologram that can be viewed from any angle.

3. *Synchronized recording*: The cameras are synchronized in such a way that they record simultaneously, ensuring that all perspectives are captured in unison. This is essential for creating a cohesive 3D model without inconsistencies.

4. *Depth-sensing technology*: In addition to standard video or image capture, these cameras often include *depth sensors* (like LiDAR or infrared cameras) to measure the distance of various points on the subject. This depth data is essential for creating an accurate 3D representation of the object or person being captured.

5. *Motion tracking*: Some systems use cameras with *motion-tracking capabilities* to capture any movement of the subject during recording, which can be used to create dynamic holograms that change or animate over time.

6. *Calibration and alignment*: To ensure that the data from all the cameras and sensors can be stitched together seamlessly, the system relies on precise calibration. Each camera's position, angle, and lens properties are calibrated to ensure that their footage aligns correctly in the post-processing phase.

These specialized systems capture raw data, which is then processed and stitched together using advanced algorithms to create a fully immersive hologram with high fidelity, accurate depth, and lifelike motion.

HoloConnects employs *advanced AI algorithms* and *machine learning techniques* to process the multi-angle image or video data collected by specialized cameras. Here is how the algorithms work in technical terms.

1. Image and video processing

   Computer Vision algorithms are applied to analyze the multi-angle images and videos. These algorithms focus on detecting key features such as edges, textures, and contours of the subject to ensure high-fidelity representation in the hologram.

   Techniques like convolutional neural networks (CNNs) are used to handle image segmentation, object recognition, and feature extraction, which are essential for understanding the spatial properties of the subject.

2. Depth synthesis and 3D reconstruction

   The depth data captured from multiple camera angles is processed using multi-view stereo (MVS) or structure from motion (SfM) algorithms. These algorithms reconstruct the 3D geometry of the subject by combining the various perspectives into a unified model.

   AI-driven algorithms assist in smoothing and filling gaps in the 3D model to create a more realistic structure. Techniques like deep learning-based 3D point cloud processing may be used to refine the depth information for accurate hologram rendering.

3. Machine learning for data synthesis

   Generative models such as Generative Adversarial Networks (GANs) or Variational Autoencoders (VAEs) can be applied to enhance visual realism by synthesizing missing data, improving texture quality, and adding finer details that might not have been fully captured during the initial recording.

   Pose estimation algorithms are often used to analyze movement and ensure that dynamic holograms (for moving subjects) maintain smooth and natural motion throughout the rendering process.

4. Depth perception and realism

   Neural rendering techniques are applied to ensure that the hologram has proper depth perception and accurately represents how light interacts with the subject from different angles. This can involve AI algorithms that simulate lighting effects and shadow placement based on the 3D model.

   To further enhance realism, recurrent neural networks (RNNs) or long short-term memory (LSTM) networks might be used to track changes in motion or lighting conditions over time, enabling lifelike holograms that adapt in real-time.

5. Data compression and optimization

   Once the hologram data is generated, it undergoes AI-based optimization to reduce the computational load required for rendering while maintaining the quality of the hologram. Techniques such as mesh simplification and texture mapping are optimized using AI to ensure efficient storage and transmission without sacrificing quality.

   These AI-driven algorithms ensure that the final hologram not only reflects the visual details of the original subject but also provides an immersive and accurate 3D experience with proper depth perception and lifelike qualities.

4.1.3 3D reconstruction

Using the processed data, the system constructs a detailed 3D model of the subject. This involves generating a mesh structure and applying textures to create a realistic appearance. The AI algorithms play a significant role in refining this model to ensure that it is as accurate and lifelike as possible.

In this context, the phrase *"mesh structure and applying textures"* refers to two important stages in building a 3D model:

- *Mesh structure*: This is the framework of the 3D model. It consists of interconnected points (vertices) that form polygons, creating the object's shape and contours in three dimensions. Imagine it as a digital skeleton or wireframe that outlines the subject's physical form.

- *Applying textures*: After the mesh structure is created, *textures* (detailed 2D images) are applied to the surface of the mesh. These textures add visual detail, such as color, patterns, or material properties (like skin or fabric), making the model look realistic.

In short, the *mesh structure* defines the shape, and *applying textures* adds surface details. AI algorithms help refine both the mesh and textures to make the model as accurate and lifelike as possible.

4.1.4 Holographic rendering

The 3D model is then rendered as a hologram. This step involves converting the 3D model into a series of light fields that can be projected to create the illusion of a three-dimensional object. HoloConnects [8] uses state-of-the-art light field technology to project these holograms. Light fields are collections of light rays captured from different angles, which, when projected, recreate the full parallax effect, allowing viewers to see the hologram from different perspectives.

4.1.5 Projection and display

The final step involves projecting the holographic image [9] using the Holobox. This display system uses an array of micro-optical elements to project light fields, creating a true 3D holographic image [9] that can be viewed without special glasses. The Holobox is capable of displaying these images with high brightness, contrast, and color fidelity, making the holograms appear vivid and realistic, A *Holobox hologram* is a display technology that creates 3D holographic visuals, allowing users to experience lifelike images that appear to float in space. Here is a detailed explanation of how this technology works.

1. Display technology

   The core of the Holobox is usually an LED screen or a high-definition projection system. This screen can be placed at an angle behind a transparent or semi-transparent surface, which is essential for creating the illusion of 3D depth.

   In some Holobox designs, the screen might use 4 K or 8 K resolution to ensure that the holographic visuals are crisp and clear.

2. Optical illusion via glass or acrylic panels

   The key element in the Holobox is a transparent, angled glass or acrylic panel (sometimes called a beam-splitter glass). This panel reflects the light from the display and allows viewers to see the reflected image as if it were floating in mid-air.

   The use of the transparent surface allows for the illusion of depth and 3D because it reflects the image while still allowing some background visibility, creating the floating hologram effect.

3. 3D content creation and projection

   The content shown in the Holobox is created using 3D modeling software or is captured via multi-angle cameras. The holographic imagery or video is designed to give the illusion of depth by carefully adjusting how the content appears on the screen, using multiple angles or layers.

   The system typically uses computer graphics to generate or process these images. Depending on the setup, the visuals may be pre-rendered 3D models or live content captured using depth-sensing cameras.

4. Holographic illusion via peppers ghost effect

   Many Holobox systems use a technique called Pepper's Ghost, which has been modernized for digital holography. This effect works by projecting or reflecting an image onto an angled glass surface, creating the illusion that the image is three-dimensional and floating in space.

   When combined with lighting adjustments and visual positioning, the effect becomes more immersive, giving viewers the impression that they are seeing a true 3D hologram.

5. AI and motion tracking (advanced Holoboxes)

   Advanced Holobox systems can integrate AI-based motion tracking. These systems use depth sensors or infrared cameras to track the viewer's position, adjusting the angle and perspective of the hologram in real time, so it appears correctly oriented no matter where the viewer stands.

   This technology can create an interactive experience where viewers can move around the box, and the hologram adjusts to maintain the 3D illusion.

6. Audio and interactivity

   Many Holobox systems are also equipped with spatial audio or directional sound systems to enhance immersion. The audio changes based on the viewer's position, further reinforcing the 3D effect.

   Some Holoboxes include touch or gesture controls, allowing users to interact with the holographic content, often using infrared sensors to detect hand motions or other gestures.

7. Final rendering and display

Once the content is processed and projected, the combination of high-resolution display, the reflective surface, and 3D software results in a hologram that appears to float in space. The illusion is reinforced by surrounding ambient lighting and precise display calibration, making the 3D model look lifelike from various viewing angles.

8. Summary of key components

   - High-resolution display (LED or projection)
   - Transparent, angled glass panel (beam-splitter or acrylic for reflection)
   - 3D content generation using modeling software or multi-angle footage
   - Pepper's Ghost effect for creating the 3D illusion
   - AI and motion tracking for perspective adjustments
   - Spatial audio and interactivity to enhance immersion.

   In essence, Holobox hologram technology combines high-resolution visual displays, optical illusion techniques, and real-time interactive elements to produce a captivating 3D experience. **Figure 1** demonstrates how this happens.

4.1.6 Interactive features

HoloConnects [8] also integrates interactive features into their holographic displays. Users can interact with the holograms using gestures or touch, facilitated by sensors and cameras that track user movements. This interaction enhances the user experience, in this context refers to how *HoloConnects* transforms a simple

Figure 1.
Holoconnects device working, the "holobox" [8].

visual display into an engaging, dynamic, and immersive interaction by allowing users to interact with the holograms. Here is a detailed breakdown of what this means.

1. Sensors and cameras for tracking user movements

   HoloConnects employs *sensors* (such as infrared or depth sensors) and *cameras* (like motion-tracking or optical sensors) to monitor the user's gestures or physical movements in real time.

   These sensors detect user actions—such as hand motions, pointing, or even touch—without the need for physical contact. This enables the system to respond to the user's gestures by altering the hologram's behavior or appearance.

2. Interactive characteristics:

   Users can *manipulate or interact* with the holographic elements, much like interacting with physical objects. For instance, they might rotate, zoom, or move parts of the hologram using simple gestures like swiping or pinching.

   Some systems may also support *voice commands* or advanced hand-tracking, allowing users to command the hologram to perform actions, such as opening menus, displaying additional information, or changing perspectives.

3. Enhanced user engagement

   This interactivity makes the hologram far more engaging compared to a static visual. The ability to manipulate and explore the hologram gives users a sense of *agency and control* over the experience, fostering a deeper connection with the content.

   For example, if a user can rotate a 3D holographic model of an object to view it from all angles or manipulate specific parts, it creates a more *immersive learning or entertainment experience*.

4. Customization and real-time interaction

   The system can *respond in real time* to a user's input, offering immediate feedback. For example, in an educational setting, a user might interact with a hologram of a molecule, pulling atoms apart to explore its structure or change its composition. This real-time response adds to the sense of immersion and makes the experience more personalized.

   By reacting to the user's actions, the hologram adapts to the user's needs, enhancing the feeling of *customized interaction*. It allows users to explore content in a way that fits their preferences and pace.

5. Physical and cognitive immersion

   When users can physically interact with a hologram, they become more *mentally and physically immersed* in the experience. This is because the interaction combines *visual, cognitive, and motor skills*, creating a richer, multi-sensory experience.

For instance, in a holographic presentation of architectural models, users might walk around the hologram to get a 360-degree view, zoom in to examine details, or even open doors and explore interiors by touching or gesturing. This type of interaction fosters *deeper learning and retention* compared to passive observation.

6. Greater engagement and entertainment

   The interactive features also make the holographic display more fun and engaging, particularly in *entertainment or gaming* contexts. The ability to directly influence the content on-screen makes users feel like they are part of the scene, turning a passive viewing experience into an *active and entertaining one*.

4.2 Detailed process of creating holograms: ASKA3D technology

ASKA3D [10] is another innovative player in the field of holographic technology, offering unique approaches to creating three-dimensional visualizations. Here is a detailed look at their process and technology.

4.2.1 Data acquisition

The creation of ASKA3D [10] holograms begins with capturing detailed images or video footage from multiple angles. This process involves using high-resolution cameras to ensure that every aspect of the subject is recorded accurately. The data collected during this phase is crucial for creating a comprehensive three-dimensional model.

The creation of *ASKA3D* holograms involves several technical steps, starting with the process of capturing detailed imagery. Here is a high-detail breakdown of how *high-resolution cameras* play a crucial role in this process.

1. Multi-angle image and video capture

   The process begins by placing the subject (whether it is a person, object, or scene) in an environment where multiple high-resolution cameras are strategically positioned. These cameras are often arranged around the subject at different angles to capture a complete 360-degree view.

   High-resolution cameras, typically capable of capturing images at 4 K, 8 K, or higher resolutions, are used to record the fine details of the subject from all perspectives. This includes textures, depth information, lighting variations, and subtle details that lower-resolution cameras might miss.

2. Synchronized recording for accuracy

   These cameras are often synchronized to capture simultaneous footage or images. Synchronized shooting ensures that all the cameras capture the subject at the same time and with the same lighting conditions. This prevents inconsistencies when stitching the images together in later stages.

   If the subject is in motion, this synchronization ensures that each frame accurately represents a real-time moment from multiple viewpoints, critical for creating dynamic, moving holograms.

3. Depth and detail collection

    - High-resolution cameras are crucial for capturing the fine details of the subject, such as textures, small movements, and lighting nuances. This level of detail is essential for ensuring that the final hologram appears realistic and lifelike.

    - Depth-sensing cameras may also be used in conjunction with standard high-resolution cameras. These specialized sensors capture additional information about the distance of the subject from the camera, which is crucial for accurate 3D modeling and hologram creation.

4. Data collection for 3D model creation

    Once the cameras capture the multi-angle footage or images, the raw data is compiled. This data includes not only the visual aspects of the subject (color, shape, texture) but also information about depth and spatial orientation.

    The high resolution of the captured images ensures that every minute detail—such as skin pores, fabric texture, or reflections—are recorded with accuracy. This data is essential when building the three-dimensional model, as it provides the base upon which the hologram will be constructed.

    In this context, "high resolution" refers to the capability of the cameras or imaging systems used to capture images with a high level of detail and clarity. Here is what high resolution specifically means in this context:

    1. Pixel density

        High resolution means the images have a high pixel density. This is measured in terms of pixels per inch (PPI) or the total number of pixels in the image (e.g., 4 K and 8 K). More pixels allow for finer details to be captured, as each pixel represents a smaller part of the image.

    2. Detail capture

        High resolution ensures that minute details of the subject are clearly captured. This includes small features like skin pores, fabric weaves, and reflections that might be missed with lower-resolution imaging. The more pixels there are, the more detailed and nuanced the captured image will be.

    3. Accuracy in representation

        When building a 3D model or hologram, high-resolution images provide a more accurate and precise representation of the original object. This is crucial for achieving a realistic hologram that mirrors the original object's appearance and texture.

    4. Clarity and sharpness

        High resolution contributes to greater clarity and sharpness in the images. This means that even the smallest features are not blurred or pixelated, resulting in a clearer and more detailed final model or hologram.

5. Data for 3D modeling

For constructing a 3D model, high-resolution images provide the base data that is used to create a detailed and accurate representation of the subject. The more detailed the initial images are, the more detailed the 3D model and hologram will be, capturing all the intricacies of the original subject.

5. Stitching and alignment

- The images or video footage from each camera are then processed to align and stitch together the various perspectives. This process typically uses sophisticated algorithms that merge the individual viewpoints into a single, cohesive 3D model.
- The high resolution of the cameras ensures that there is enough detail in the overlapping areas between different camera views, allowing for smooth transitions and reducing artifacts or distortions in the final hologram.

6. Creating the 3D mesh

After the data is processed, a 3D mesh is constructed. The mesh forms the geometric framework of the subject, defining its shape and contours in 3D space. The high resolution of the initial imagery ensures that the mesh is highly detailed, capturing even the smallest elements of the subject's form.

The clarity provided by the high-resolution data allows for more precise modeling, which is particularly important when rendering realistic holograms.

7. Applying textures and finalizing the hologram

Once the mesh is created, textures are applied to it. The high-resolution images or video footage serve as the source for these textures, adding realistic surface details to the model (such as skin tone, clothing fabric, or material reflections).

The accuracy of these textures depends directly on the quality of the initial images captured by the high-resolution cameras. Without high-quality source data, the hologram would lack the fine detail that gives it a realistic appearance.

4.2.2 Image processing and conversion

Once the data is collected, ASKA3D [10] employs advanced image processing algorithms to convert these images into a three-dimensional holographic format. This involves analyzing the captured images to determine depth information and reconstructing the 3D structure of the subject. The conversion process ensures that the resulting hologram accurately represents the original object from all angles.

Accuracy refers to how closely a measurement or representation matches the true or real value of what is being measured. In the context of *holography or 3D modeling, accuracy* relates to how precisely the hologram or model replicates the physical characteristics of the original object, including its:

1. *Shape*: The hologram's 3D structure should accurately reflect the contours and geometry of the original object.

2. *Proportions*: The relative dimensions, sizes, and spatial relationships in the hologram must match the real-world object.

3. *Details*: Fine textures, patterns, or surface features must be represented without distortion or loss of quality.

4. *Depth*: The hologram should represent accurate depth perception, meaning it maintains the correct spatial arrangement as the original object when viewed from different angles.

In summary, *accuracy* in this context refers to the hologram's ability to faithfully reproduce the original object in a way that is indistinguishable from the real thing when it comes to these visual and structural aspects.

This refers to a set of complex computational techniques used to convert two-dimensional (2D) image data into a three-dimensional (3D) model, as required for the creation of a hologram. In the context of *ASKA3D*, these algorithms perform several specialized tasks to accurately reconstruct the 3D structure of a subject from the collected image data.

Here is a detailed breakdown of the types of algorithms likely involved in this process:

1. Multi-view stereo (MVS) algorithms

    - Multi-view stereo algorithms take the multiple 2D images captured from various angles and match corresponding points between the images. By comparing how the subject appears in different images, the algorithm estimates the *depth information* and reconstructs the 3D geometry of the subject.

    - These algorithms analyze the differences between the camera perspectives to extract depth, producing a *3D point cloud*—a set of data points representing the subject's surface in 3D space.

2. Structure from motion (SfM) algorithms

    Structure from Motion algorithms are often used to reconstruct 3D models from 2D images when the subject or camera is in motion. The SfM technique uses changes in perspective to calculate both the relative positions of the cameras and the 3D structure of the subject.

    This type of algorithm is essential in reconstructing 3D models from image sequences, making it valuable for holographic displays where *dynamic movement* is involved.

3. Depth estimation algorithms

    - Depth estimation algorithms work by analyzing the parallax (apparent shift in position) between multiple images. By understanding how different parts of the subject shift between perspectives, the algorithm calculates how far each point is from the camera.

- *Depth maps* are generated, which are essential for determining the *3D structure* of the subject. This information is critical for ensuring that the hologram has realistic depth and perspective when viewed from different angles.

4. Photogrammetry

- Photogrammetry involves measuring the distance between objects in the captured images to reconstruct the subject's 3D shape. The advanced photogrammetric algorithms match common features between images and use these points to estimate the geometry of the subject.

- These algorithms help ensure that the *3D model accurately replicates the proportions* and spatial relationships of the original object, contributing to a highly detailed hologram.

5. Surface reconstruction algorithms

- Once depth and point cloud data are available, *surface reconstruction algorithms* are employed to create a smooth, continuous surface over the 3D point cloud. These algorithms connect the points to form a mesh or surface that represents the subject's structure.

- This step transforms raw data into a usable 3D model, ensuring that the hologram displays a solid, visually coherent subject rather than just a collection of points.

6. Texture mapping and image stitching

To enhance the realism of the 3D model, texture mapping algorithms apply the original high-resolution images (captured by cameras) onto the surface of the 3D mesh. These algorithms ensure that the *textures (colors, details, and patterns)* from the 2D images are properly aligned with the 3D structure.

Image stitching algorithms are used to seamlessly blend the visual data from different camera angles, ensuring a consistent appearance across the entire 3D model.

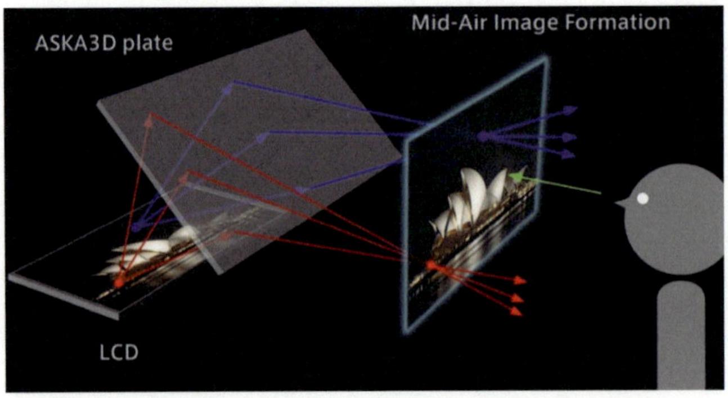

Figure 2.
ASKA3D hologram plate how it works [10].

7. Neural networks for super-resolution and refinement

   Some advanced image processing algorithms use *neural networks*, particularly in cases where the raw data might need additional refinement. AI-driven algorithms like *convolutional neural networks (CNNs)* are often employed to enhance the *resolution of textures* or to fill in gaps in the 3D model.

   These networks can predict missing details based on the surrounding visual data, resulting in a more *realistic and detailed hologram*.

4.2.3 Light field rendering

ASKA3D [10] uses a technology known as light field rendering to generate holograms. This technique involves simulating how light interacts with the object from various angles, creating a light field that can be projected to form a 3D image. The light field is essentially a collection of light rays that carry information about the direction, intensity, and color of light as it interacts with the object.

4.2.4 Holographic display technology

The heart of ASKA3D's [10] technology lies in its proprietary holographic display panels. These panels are designed with an array of micro-optical elements that control the direction of light rays. By manipulating these light rays, the display can project a three-dimensional image that appears to float in mid-air. This method allows viewers to see a realistic 3D image without the need for special glasses, as shown in **Figure 2**.

4.2.5 Projection and interaction

ASKA3D's [10] displays are capable of projecting high-resolution holograms that can be viewed from different angles, providing a true 3D experience. The displays are also equipped with sensors to detect viewer position and movement, enabling the hologram to adjust dynamically for an interactive experience. This capability is particularly useful in applications such as virtual meetings, where participants can interact with holograms in real time.

4.2.6 Applications of ASKA3D holograms

Entertainment: ASKA3D [10] holograms are used in concerts, exhibitions, and theme parks to create immersive experiences.

Healthcare: They provide detailed 3D visualizations for medical training and surgical planning.

Education: ASKA3D [10] holograms enhance learning by providing interactive 3D models that make complex concepts easier to understand.

Retail: Holograms are used to create engaging displays and advertisements, attracting customers with eye-catching 3D visuals.

ASKA3D [10] enables touchless interaction for security by using mid-air holographic displays that detect user gestures, allowing secure input without physical contact, thus eliminating the risk of leaving traces.

4.3 Light Field Lab: Detailed process of creating holograms

Another pioneering company, Light Field Lab [11], focuses on creating SolidLight [12] holographic displays, which produce high-resolution, full-color holograms that do not require any special glasses. SolidLight [12] technology uses complex light field arrays to project 3D images into mid-air, providing an unparalleled visual experience. Light Field Lab's [11] displays can create holograms that appear as real as physical objects, offering significant potential for applications in entertainment [5], advertising, and more. The company aims to integrate this technology into various platforms, enabling immersive storytelling and interactive experiences that were previously the domain of science fiction.

Holographic technology is revolutionizing various sectors, and recent advances in holographic displays are spearheaded by innovative companies like HoloConnects [8] and Light Field Lab [11]. These companies are developing astonishing 3D visual holograms that are transforming the way we experience digital interactions.

Light Field Lab [11] is at the forefront of developing advanced holographic displays. Their innovative approach leverages light field technology to create stunning, high-resolution holograms that can be viewed without the need for special glasses. Here is a detailed look at their process.

4.3.1 Data collection and input

Light Field Lab [11] starts by capturing high-resolution images or videos from multiple angles around the subject. This is similar to the initial step in HoloConnects' [8] process but focuses on creating a more immersive and realistic holographic experience. The data collection phase is critical for ensuring that all perspectives of the subject are accurately recorded.

4.3.2 Light field generation

Once the images are collected, they are processed to generate light fields. A light field is a representation of the amount of light traveling in every direction through every point in space. Light Field Lab [11] uses sophisticated algorithms to convert the

Figure 3.
Light Field Lab hologram module projection [11].

Figure 4.
Light Field Lab's interaction hologram example [11].

captured data into a comprehensive set of light fields, effectively mapping how light interacts with the subject from different angles.

4.3.3 Holographic display technology

The core of Light Field Lab's [11] innovation lies in their holographic display technology. This involves a complex array of micro-optical elements, including lenses and mirrors, that project light fields. These elements are meticulously aligned to ensure that the light fields combine to form a coherent three-dimensional image that appears to float in mid-air.

4.3.4 Projection and viewing

The generated light fields are then projected using Light Field Lab's [11] proprietary display panels. These panels consist of millions of tiny lenses that direct light rays in specific directions. As a result, viewers can see a 3D image with full parallax, meaning the image shifts naturally as the viewer moves around it, offering a lifelike holographic experience, in **Figure 3** shows an example of that.

4.3.5 Real-time interaction

Light Field Lab's [11] displays are designed to support real-time interaction. This is achieved by integrating sensors that track the viewer's position and movements, allowing the hologram to adjust dynamically. This interaction is crucial for applications in various fields, from industrial work to entertainment [5], the same to education [13] and medical visualization, as seen in **Figure 4**.

4.3.6 High-resolution imaging

One of the standout features of Light Field Lab's [11] technology is its ability to produce holograms with extremely high resolution. This is accomplished through their advanced light field rendering algorithms, which ensure that the projected images are sharp, detailed, and vivid.

4.3.7 Applications and use cases

Light Field Lab's [11] holographic displays have a wide range of applications:

- *Entertainment*: Creating immersive experiences in theme parks, concerts, and virtual reality.
- *Communication*: Enhancing telepresence with realistic, 3D representations of participants.
- *Medical field*: Offering detailed 3D visualizations of medical data for diagnostics and training.
- *Education*: Enabling interactive and engaging learning experiences through lifelike 3D models.

5. Transformative impact of holographic technology

The impact of holographic technology is being felt across a multitude of sectors. In the entertainment [5] industry, holography is revolutionizing the way audiences experience concerts, movies, and video games, offering immersive experiences that were previously unimaginable. HoloConnects [8] and Light Field Lab's [11] advancements are setting new standards in this field, providing more realistic and engaging visuals.

In communication [6], holographic displays are enabling more engaging and lifelike interactions, potentially transforming virtual meetings and teleconferences. Companies like HoloConnects [8] are enhancing business meetings and educational sessions by creating shared holographic spaces that foster a sense of presence and collaboration.

Healthcare [7] is another field where holography is making significant strides. Advanced holographic imaging techniques are providing doctors with detailed, three-dimensional views of patient anatomy, enhancing diagnostic accuracy and improving surgical planning. For example, HoloConnects' [8] technology allows for real-time, live holographic consultations, bridging the gap between doctors and patients in remote locations.

Furthermore, holography is being used in educational settings to create interactive learning environments, helping students grasp complex concepts through visual and experiential learning. The use of holograms in education [13] can make abstract subjects more tangible and understandable, thereby enhancing the overall learning experience.

These examples highlight the transformative potential of holographic technology across various industries, driven by the pioneering efforts of companies like HoloConnects [8] and Light Field Lab [11]. Their innovations are paving the way for a future where holograms become a ubiquitous part of our daily lives, enriching how we protect, communicate, learn, and entertain ourselves.

For more detailed information about their technologies and to see their holograms in action, you can visit the official websites of HoloConnects [8] and Light Field Lab

[11]. These sites provide comprehensive insights into their holographic solutions and showcase stunning visuals of their latest advancements.

5.1 Future prospects and societal implications

As holographic technology continues to evolve, its future prospects appear increasingly promising. The development of more advanced and affordable holographic displays will likely lead to broader adoption across various domains. This widespread adoption could have profound societal implications, altering the way we interact with digital content and each other.

Holography holds the potential to enhance not only entertainment [5] and communication [6] but also education [13], healthcare [7], and numerous other fields. By providing more intuitive and immersive ways to visualize information, holography can improve our understanding of complex data and facilitate better decision-making.

5.2 Artificial intelligence (AI) in hologram development

5.2.1 Data acquisition and preprocessing

AI [14] algorithms play a crucial role in the initial stages of hologram development by enhancing data acquisition and preprocessing.

Computer vision and machine learning: AI-driven computer vision techniques [14] are employed to capture and analyze images from multiple angles. Machine learning models can automatically detect and correct errors in the captured images, such as noise reduction and artifact removal, ensuring high-quality input data.

5.2.2 3D reconstruction

AI [14] algorithms, particularly deep learning models, can reconstruct 3D models from 2D images. Convolutional neural networks (CNNs) are used to infer depth information and create accurate 3D representations of objects from multiple 2D perspectives.

5.2.3 Hologram generation

AI [14] techniques are integral to the complex process of generating holograms from the acquired data.

Generative adversarial networks (GANs): GANs can generate high-resolution holograms by learning from a dataset of existing holograms. The generator network creates holograms, while the discriminator network evaluates their quality, leading to continuous improvement through adversarial training.

Phase retrieval algorithms: AI [14] improves phase retrieval algorithms, which are essential for converting intensity measurements into phase information. Neural networks can optimize these algorithms to produce more accurate phase maps, which are crucial for realistic hologram rendering.

5.2.4 Light field rendering

Light field rendering, essential for holographic displays, benefits significantly from AI [14].

Light field networks: AI-driven light field networks [14] can simulate the way light interacts with objects. These networks use deep learning to predict light behavior, ensuring realistic rendering of holograms. They can handle complex light interactions such as refraction, reflection, and scattering.

Volumetric rendering: AI [14] enhances volumetric rendering techniques, which involve creating 3D images from volumetric data. Machine learning models can optimize the rendering process, making it faster and more efficient while maintaining high visual fidelity.

5.2.5 Real-time holographic display

AI [14] enables real-time processing and display of holograms, which is critical for interactive applications.

Neural rendering: Neural rendering techniques use AI [14] to render holograms in real-time. These techniques leverage neural networks to predict and render images quickly, enabling real-time interaction with holographic content.

Edge computing: AI-driven edge computing [14] allows for decentralized processing of holographic data, reducing latency and enhancing real-time performance. AI algorithms running on edge devices process and render holograms locally, ensuring a smooth user experience.

5.2.6 Optimization and enhancement

AI [14] optimizes various aspects of holographic technology, enhancing both the creation process and the quality of the final product.

Algorithm optimization: AI algorithms [14] can optimize hologram generation and rendering processes. Reinforcement learning models, for example, can find the most efficient ways to compute holograms, reducing computational load and increasing speed.

Super-resolution: AI [14] techniques such as super-resolution algorithms enhance the resolution of holograms. These algorithms can predict and fill in missing details, resulting in sharper and more detailed holographic images.

Figure 5.
Holographic Keypad from ASKA3D technologies [10].

5.2.7 Enhancing security with holographic technology

Holographic technology offers a futuristic approach to enhancing the security of ATM transactions, specifically by enabling users to enter security keys without leaving any traceable marks. This technology leverages advanced holographic displays and AI-driven systems [14] to create secure and private interactions between users and ATMs. Here is how holograms work in this context and improve security.

5.3 Holographic keypads

Holographic keypads project a virtual keyboard in front of the user, allowing them to enter their PIN or password without physically touching any surface. Here is how it works.

Projection: A small projector within the ATM generates a 3D holographic image of a keypad in the air.

Interaction: Users interact with the holographic keypad by hovering their fingers over the virtual keys. Advanced sensors detect the position and movement of the fingers, interpreting them as key presses.

No physical contact: Since there is no physical contact with a screen or keypad, there are no fingerprints or smudge marks left behind, significantly reducing the risk of someone deciphering the entered password, in **Figure 5**, you can see an example of how this technology works in real life.

5.4 Advanced sensors and AI integration

The integration of AI [14] and advanced sensors enhances the security and accuracy of holographic keypads:

3D motion tracking: AI-driven sensors [14] track the precise movement of the user's fingers in three dimensions. This ensures accurate detection of key presses and prevents false inputs.

Behavioral analysis: AI systems [14] analyze user behavior patterns, such as the speed and pressure of key presses, to detect any anomalies that might indicate fraudulent activity.

Real-time adjustment: The system can adjust the holographic display in real-time based on user interactions, ensuring a seamless and intuitive experience.

5.5 Secure data transmission

Holographic systems ensure that the data entered via the holographic keypad is transmitted securely:

Encryption: All data entered through the holographic keypad is encrypted immediately, ensuring that it cannot be intercepted or accessed by unauthorized parties.

Isolation: The holographic input system operates in a secure environment, isolated from potential malware or hacking attempts that might affect traditional touchscreens.

5.6 User privacy

Holographic technology also enhances user privacy, which is crucial for ATM transactions:

Invisible input: Because the holographic display is only visible to the user directly in front of the ATM, it is difficult for onlookers or cameras to capture the input. This prevents shoulder surfing and other forms of spying.

Dynamic keypad layouts: The layout of the holographic keypad can change dynamically, making it harder for malicious actors to learn the layout and predict the user's key presses.

6. Conclusion

Holographic technology, once a staple of science fiction, has undergone significant advancements and is now on the cusp of transforming numerous industries and everyday life. Through the concerted efforts of innovative companies like Holoconnects [8] and Light Field Lab [11], along with the application of artificial intelligence [14], the potential of holography is being fully realized, pushing the boundaries of visual communication and immersive experiences.

6.1 Evolution of holographic technology

From its humble beginnings rooted in scientific experimentation, holography has evolved into a multifaceted technology with diverse applications. Initially, the focus was on simple 3D imaging, but over the years, the technology has expanded to include complex light field displays, real-time rendering, and interactive environments. The transition from traditional 2D displays to dynamic 3D holograms represents a major leap forward in how we visualize and interact with information.

6.2 AI-driven advancements

AI [14] plays a pivotal role in enhancing the capabilities of holographic technology. AI techniques such as machine learning, deep learning, and neural networks are crucial for:

Data acquisition and preprocessing: AI [14] enhances the quality of input data by reducing noise and correcting errors, ensuring high-fidelity holograms.

Hologram generation: AI-driven algorithms [14], including GANs and phase retrieval models, improve the accuracy and realism of holograms.

Light field rendering: AI [14] enables realistic light field simulations, essential for creating lifelike holographic displays.

Real-time processing: Neural rendering and edge computing powered by AI allow for real-time interaction with holographic content.

Optimization: AI [14] optimizes computational processes, making hologram generation and rendering more efficient.

6.3 Contributions from leading companies

Holoconnects: This company utilizes deep learning algorithms to enhance 3D reconstruction accuracy and optimize light field rendering. Their AI-driven approach [14] results in more realistic and interactive holographic experiences, demonstrating the practical application of AI [14] in improving holographic technology.

Light Field Lab: Specializing in true holographic displays, Light Field Lab [11] employs neural networks to simulate light interactions and volumetric rendering techniques to produce high-fidelity holograms. Their technology exemplifies the cutting-edge advancements made possible by integrating AI [14] with holography, pushing the envelope of visual realism and depth.

ASKA3D: Utilizing AI [14], ASKA3D [10] focuses on refining the hologram generation process. By leveraging AI [14] for phase retrieval and light field simulations, they produce highly detailed and accurate holograms, contributing to the broader adoption of holographic technology in various fields.

6.4 Impact across industries

The impact of these advancements is being felt across multiple sectors:

Entertainment: Holography is revolutionizing concerts, movies, and video games, offering immersive experiences that engage audiences in unprecedented ways.

Communication: Holographic displays enhance virtual meetings and teleconferences, providing more engaging and lifelike interactions.

Healthcare: Advanced holographic imaging techniques offer doctors detailed, three-dimensional views of patient anatomy, improving diagnostic accuracy and surgical planning.

Education: Holography creates interactive learning environments, helping students understand complex concepts through visual and experiential learning.

Industrial uses: With no leaving traces in a surface, the projection hologram is very useful to apply into ATM's, as well as restricted access panels and different controls, like elevators.

6.5 Future prospects

The integration of AI [14] with holographic technology marks a significant milestone in the evolution of visual communication. The advancements made by companies like Holoconnects [8], Light Field Lab [11], and ASKA3D [10] highlight the potential of AI [14] to enhance the creation, rendering, and real-time interaction of holograms. As these technologies continue to develop, we can expect even more sophisticated and immersive holographic experiences, further blurring the line between virtual and reality.

In conclusion, the synergy between AI [14] and holographic technology promises to reshape various industries, making science fiction a reality and opening up new avenues for innovation and interaction in the digital age.

Acknowledgements

In the preparation of this manuscript, I have made use of AI-driven tools to aid in refining certain aspects of the text, including assisting with phrasing and identifying grammatical errors (chagpt). These tools were employed to ensure clarity, coherence, and precision in my writing. However, it is important to note that the core content, ideas, and research presented in this work are entirely based on my original findings and analyses, which have been the subject of previous publications. The technology discussed in this paper, as well as the insights and conclusions drawn, are the product of my own research.

Author details

Camilo Fernandez
University of Antioquia, Medellin, Colombia

*Address all correspondence to: camilofernandezzapata@gmail.com

References

[1] Gabor D. Nobel lecture: Holography, 1948-1971. 1971. Nobelprize.org. Retrieved from Nobel Prize

[2] Maiman TH. Stimulated optical radiation in ruby lasers. Nature. 1960;**187**(4736):493-494. DOI: 10.1038/187493a0

[3] Hariharan P. Optical Holography: Principles, Techniques and Applications. Cambridge, UK: Cambridge University Press; 1996

[4] Benton SA, Bove VM. Holographic Imaging. Hoboken, New Jersey, USA: Wiley; 2008

[5] Holography in entertainment: Transforming audiences' experiences. Entertainment Weekly, May 5, 2024. Available from: https://www.entertainmentweekly.com/holography-in-entertainment-transforming-audiences-experiences/ [Accessed: June 19, 2024]

[6] Vance Ashlee. Advances in holographic communication: enhancing virtual meetings. Wired, February 28, 2024. Available from: https://www.wired.com/2024/02/advances-in-holographic-communication-enhancing-virtual-meetings/ [Accessed: June 19, 2024]

[7] Holography in healthcare: Improving diagnostic accuracy and surgical planning. HealthTech, April 10, 2024. Available from: https://www.healthtech.com/holography-in-healthcare-improving-diagnostic-accuracy-and-surgical-planning/ [Accessed: June 19, 2024]

[8] HoloConnects. HoloConnects. Available from: https://www.holoconnects.com [Accessed: June 19, 2024]

[9] Heath Alex. HoloConnects' AI-powered holographic displays are transforming industries. TechCrunch, March 15, 2024. Available from: https://techcrunch.com/2024/03/15/holoconnects-ai-powered-holographic-displays-are-transforming-industries/ [Accessed: June 19, 2024]

[10] ASKA3D. (n.d.). ASKA3D holographic display technology. Retrieved from https://aska3d.com

[11] Light Field Lab. Light Field Lab. Available from: https://lightfieldlab.com [Accessed: June 19, 2024]

[12] Gartenberg Chaim. Light Field Lab's SolidLight holographic displays: The future of visual experience. The Verge, April 22, 2024. Available from: https://www.theverge.com/2024/4/22/solidlight-holographic-displays-the-future-of-visual-experience/ [Accessed: June 19, 2024]

[13] Lee S, Chen H. Enhancing education through holographic simulations and artificial intelligence. Journal of Educational Technology Research. 2023;**21**(2):78-91

[14] Smith JD, Johnson AB. Artificial intelligence applications in holographic displays. Journal of Holographic Technology. 2023;**8**(2):45-56

Chapter 2

Metasurface Holography

Hui Gao

Abstract

Holography has emerged as an ideal method for future immersive naked-eye displays, as it possesses the capability to record and reconstruct light wavefronts. However, traditional dynamic holographic devices always have some shortcomings, such as a small field of view, twin imaging, and multiple orders of diffraction, which bring many challenges to their practical application. Metasurfaces, composed of subwavelength unit cells, offer great potential in manipulating light fields and addressing these limitations. The latest research progress in the field of dynamic metasurface holography is summarized from the implementation methods to design strategies, mainly including the typical research works on dynamic meta-holography based on tunable metasurfaces and multiplexed metasurfaces. The emerging applications of dynamic meta-holography have the potential to accelerate the development of light field manipulation and micro−/nanofabrication in higher dimensions. This chapter also explores potential applications and development paths in the field.

Keywords: metasurface, holographic display, dynamic meta-holography, tunable meta-holography, multiplexed meta-holography

1. Introduction

Holography, which is able to record and reconstruct wavefronts of light, enables immersive naked-eye displays and a broad range of optical applications, including optical storage [1, 2], optical encryption [3, 4], optical information processing [5, 6], and optical manipulation [7, 8]. As a powerful platform for enhancing visual immersion and realism, holography has the potential to revolutionize human-computer interaction and interpersonal communication. However, traditional optical holography relies on a complex recording process that captures the interference pattern formed by objects and reference beams [9], making it impractical for generating holographic reconstructions of virtual objects. To overcome this limitation, Brown and Lohman introduced computer-generated holography (CGH) in 1966, utilizing physical optics principles to generate interference patterns computationally [10]. CGH further enables dynamic holographic displays by integrating digital light field modulators, such as spatial light modulators (SLMs) and digital micromirror devices (DMDs) [11, 12]. Despite these advances, current holographic technologies still face challenges, including large pixel sizes and constrained modulation principles, which limit their development in areas such as field of view (FOV), twin-image suppression, bandwidth, and diffraction order control [13, 14].

In recent years, the rapid advancement of nanofabrication has spurred growing interest in metasurfaces—subwavelength nanostructured arrays capable of precise manipulation of light's amplitude, phase, and polarization [15–17]. This has led to breakthroughs in beam splitting [18, 19], metalenses [20–25], orbital angular momentum (OAM) devices [26–31], and structural-color elements [32–36]. Metasurfaces offer a compelling approach for holographic displays due to their superior light-field modulation capabilities, providing significantly more degrees of freedom for hologram design compared to conventional CGH devices. In meta-holography, target CGH patterns for holographic reconstructions are determined through physical and mathematical modeling and then implemented *via* the spatial arrangement of nanostructures. Compared to conventional CGH, meta-holography offers several advantages, including higher spatial resolution, reduced noise, broader operational bandwidth, and the suppression of undesired diffraction orders [13, 14, 37].

Meta-holography can be categorized based on different classification criteria. One common approach is to classify it by modulation type, leading to three primary categories: phase-only holography [38–42], amplitude-only holography [43–45], and complex amplitude holography (**Figure 1**) [46–49].

Additionally, meta-holography can be categorized into static and dynamic meta-holography based on the number of optical images reconstructed from a single metasurface element [50]. While static meta-holography generates a fixed image, dynamic meta-holography enables the reconstruction of multiple images from the

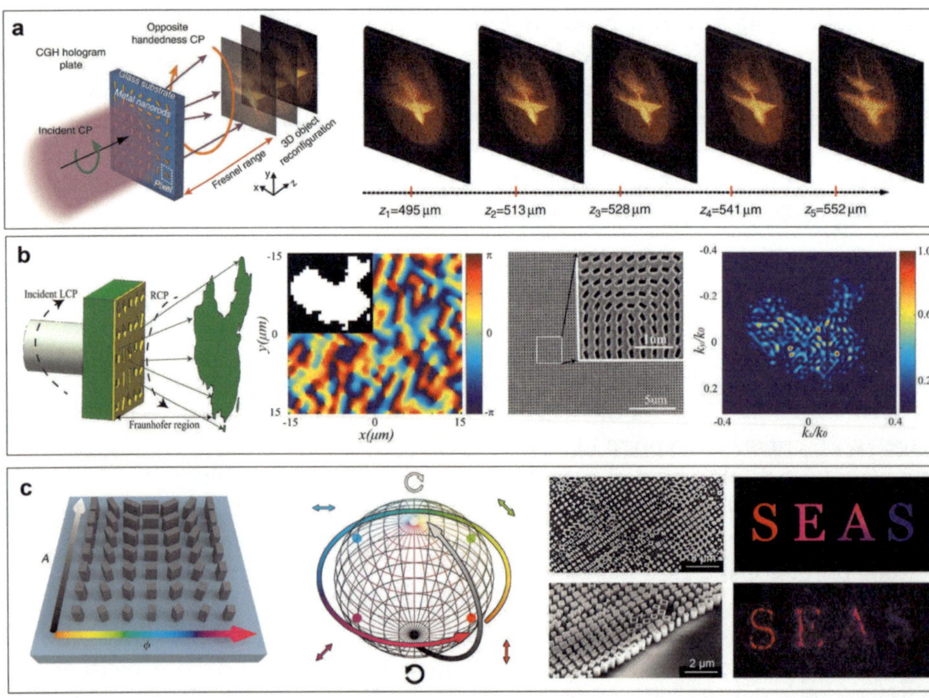

Figure 1.
Representative examples of phase-only, amplitude-only, and complex amplitude meta-holography. (a) A 3D on-axis transmission-type phase-only meta-hologram fabricated from gold nanorod arrays [39]. (b) A broadband phase-only metasurface hologram with high tolerance to fabrication errors, featuring an elongated nanoaperture array [41]. (c) A dielectric metasurface that allows complete and independent control over both optical amplitude and phase by adjusting the geometrical parameters and orientation of the meta-atoms. The central images show scanning electron microscopy (SEM) images of the fabricated samples. The experimental reconstructions are overlaid with separately measured images at wavelengths of 1.65 μm (highlighted in red) and 0.94 μm (highlighted in blue) [46]. Figure reproduced from: Refs. [39, 41, 46], Springer Nature.

same metasurface, making it more suitable for applications such as optical displays and information processing.

Several innovative meta-hologram designs have been proposed in recent years [50–53]. This chapter provides a comprehensive review of recent advancements in dynamic meta-holography, highlighting its significance and potential. Based on their implementation approaches, dynamic meta-holography can be broadly classified into two categories: tunable metasurfaces and multiplexed metasurfaces. We examine these strategies, introduce key research developments, and explore potential applications and future directions in this field.

2. Tunable meta-holography

After fabrication, most metasurfaces remain static and lack tunability. However, due to the growing demand for applications requiring dynamic control, significant efforts have been dedicated to exploring active materials and tuning mechanisms [54, 55]. Various approaches have been investigated, including thermo-optic effects (e.g., silicon [56–58], lead telluride [59]), free-carrier effects (e.g., silicon [60–62], gallium arsenide [63–65], indium antimonide [66], indium tin oxide [67–70], graphene [71–73]), phase transitions (e.g., liquid crystals [74, 75], germanium–antimony–tellurium (GST) [76–82], vanadium dioxide (VO_2) [83–86]), stretchable structures [87–90], chemical reactions [91, 92], and so on.

2.1 Phase transitions

In recent years, chalcogenide phase change materials (PCMs) based on germanium-antimony-tellurium (GST) have gained significant attention in commercial optical storage applications due to their ability to transition between amorphous and crystalline states when subjected to thermal, optical, or electrical stimuli. GST is widely utilized in dynamic meta-holography and metasurface research, benefiting from its substantial refractive index contrast, rapid switching speed, and reliable retention [76–82]. Lee et al. demonstrated the ITO-GST-ITO (IGI) meta-hologram panel concept through both theoretical and experimental approaches, achieving localized crystallization of the GST film with a large refractive index contrast using an excimer pulse laser at an impressive resolution of 16 × 16 k [79]. The fabrication of the GST hologram panel involved transferring the CGH pattern from a Cr mask to an IGI layer (**Figure 2a**). By integrating plasmonic metasurfaces with PCMs, Zhang et al. successfully demonstrated a switchable spin-orbit interaction (SOI) [80]. As a proof of concept, they fabricated and measured three reflective meta-devices that enabled the spin Hall effect, vortex beam generation, and a hologram in the amorphous state of GST ("ON" state), all of which disappeared when the GST transitioned to its crystalline state ("OFF" state) (**Figure 2b**). Zhou et al. proposed a similar approach for achieving transmissive dynamic meta-holographic encryption, designing a structure with a GST film sandwiched between a bottom SiO_2 substrate and top gold split-ring resonator (SRR) arrays [78]. The resonance characteristics of the SRR structure varied depending on the GST state, leading to distinct phase and amplitude modulations. In the amorphous state of GST, continuous phase modulation enabled holographic reconstruction, which transformed into a binary phase in the crystalline state (**Figure 2c**).

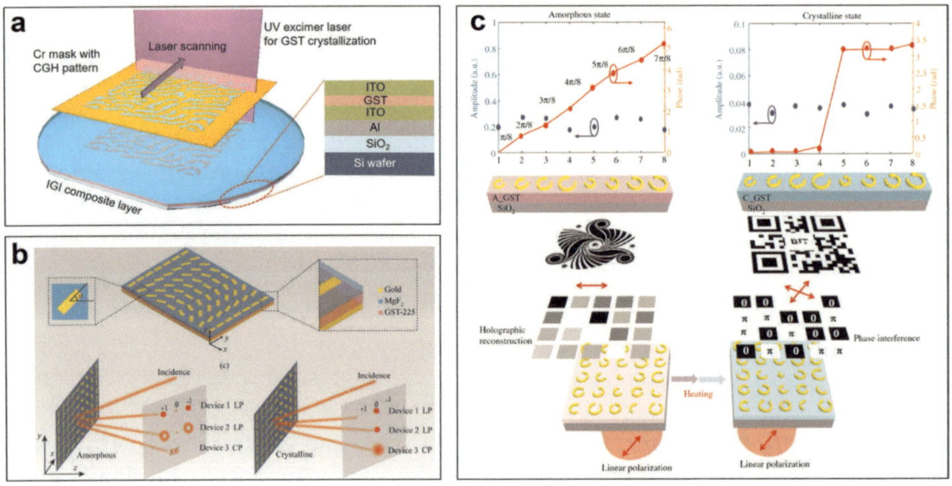

Figure 2.
Tunable metasurface holography utilizing phase transitions. (a) Schematic of a multilayered GST-based metasurface and the hologram recording process. (b) Illustration of switchable spin-orbit interactions (SOIs) demonstrated in the system. (c) Diagram depicting dual functionalities enabled by simultaneous phase modulation and interference techniques. Figure reproduced from: (a) Ref. [79], Springer Nature. (b) Ref. [80], WILEY-VCH Verlag GmbH & Co. KGaA, Weinheim. (c) Ref. [78], Walter de Gruyter GmbH.

VO_2 is another promising material for tunable meta-holography due to its insulator-to-metal transition (IMT) under various stimuli and its relatively low transition temperature (~340 K, 67°C). Driscoll et al. designed the first VO_2-based tunable metasurface device in 2008, achieving dynamic tuning of an infrared hybrid-metasurface resonance [83]. More recently, Liu et al. developed a thermally responsive dynamic meta-holography system using VO_2-integrated metasurfaces. Their design enabled the transformation of reconstructed images from "H" to "G" as the temperature increased from 25 to 100°C (**Figure 2d**) [85].

2.2 Chemical reaction

Specialized chemical reactions have also been explored for dynamic control in meta-holography. For instance, magnesium (Mg) exhibits excellent plasmonic properties in the visible spectrum and undergoes a reversible metal-to-dielectric phase transition upon reacting with hydrogen to form magnesium hydride (MgH_2) [93]. This transition can be reversed through dehydrogenation using oxygen.

Li et al. developed dynamic meta-holographic devices utilizing plasmonic Mg nanorods as addressable pixels controlled *via* hydrogenation and dehydrogenation [92]. They introduced three different types of nanorods to demonstrate applications in optical information processing and encryption. Gold (Au) nanorods (marked as P_2 in Ref. [92] and **Figure 3a**) remained chemically stable and were unaffected by the hydrogenation process. In contrast, Mg nanorods capped with a titanium (Ti) spacer and a palladium (Pd) catalytic layer (Mg/Pd, marked as P_1 in Ref. [92]) exhibited active hydrogenation and dehydrogenation behavior. Additionally, a chromium (Cr) capping layer was introduced to slow down both the hydrogenation and dehydrogenation rates in another variant (Mg/Pd/Cr, marked as P_3 in Ref. [92]). Each unit cell contained one of the three types of nanorods, uniquely encoded to generate specific holographic images (**Figure 3a**).

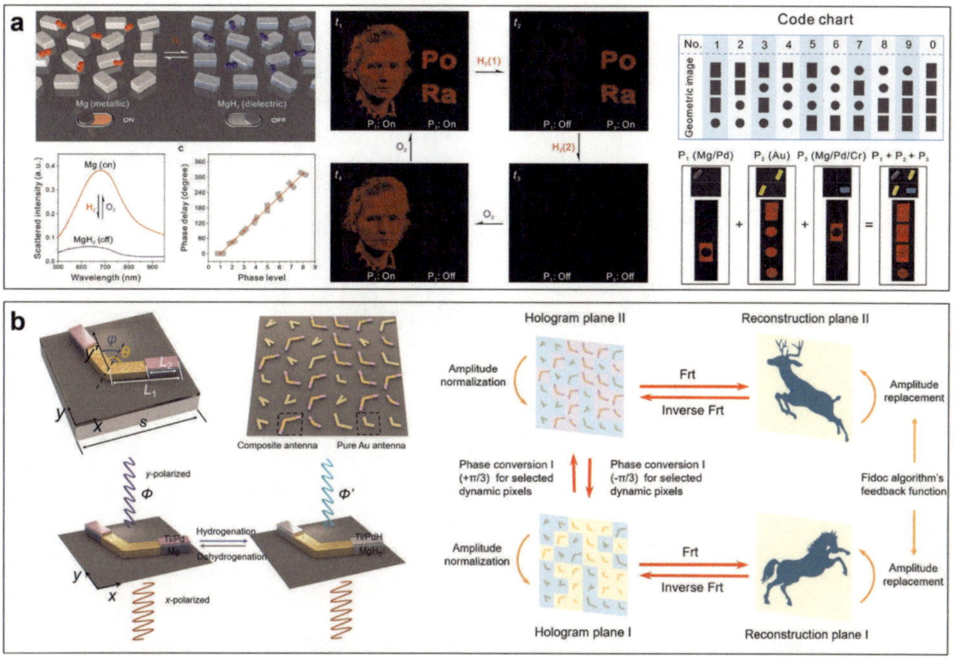

Figure 3.
Tunable metasurface holography driven by chemical reactions. (a) Chemically active metasurfaces enable independent control of addressable subwavelength pixels at visible frequencies. Plasmonic nanorods, designed to exhibit hierarchical reaction kinetics during hydrogenation and dehydrogenation, function as addressable pixels in multiplexed metasurfaces. (b) An iterative hologram algorithm based on the Fidoc method establishes a quantified phase relationship, enabling the reconfigurable phase profile to reshape the reconstructed image dynamically. Figure reproduced from: (a) Ref. [92], American Association for the Advancement of Science. (b) Ref. [91], Optical Society of America.

Furthermore, another study by Li et al. proposed a theoretical framework for an addressable dynamic meta-hologram utilizing a similar hydrogenation/dehydrogenation process with Mg [91]. Their meta-holographic devices consisted of pure gold V-shaped antennas (PVAs) and composite Mg-Au V-shaped antennas (CVAs). The scattering phase of the CVAs exhibited a uniform shift of $\pi/3$ before and after hydrogenation. To determine the optimal distribution of PVAs and CVAs, the authors developed an iterative hologram algorithm based on the Fidoc method, establishing a quantified phase relationship for precise holographic image reconstruction (**Figure 3b**).

However, chemical reaction-based methods can only enable a limited number of switchable holographic images due to the slow hydrogenation and dehydrogenation rates in these designs, which hinder smooth holographic display. As a result, chemical control methods are more suitable for platforms requiring simple dynamic functionalities without the need for rapid response, such as optical encryption and smart sensors.

2.3 Rewriting meta-holography and stretchable substrate meta-holography

As a two-dimensional material with outstanding electronic and optical properties, graphene is well-suited for achieving tunability in metasurfaces. Li et al. designed write-once holograms capable of displaying wide-angle, full-color three-dimensional

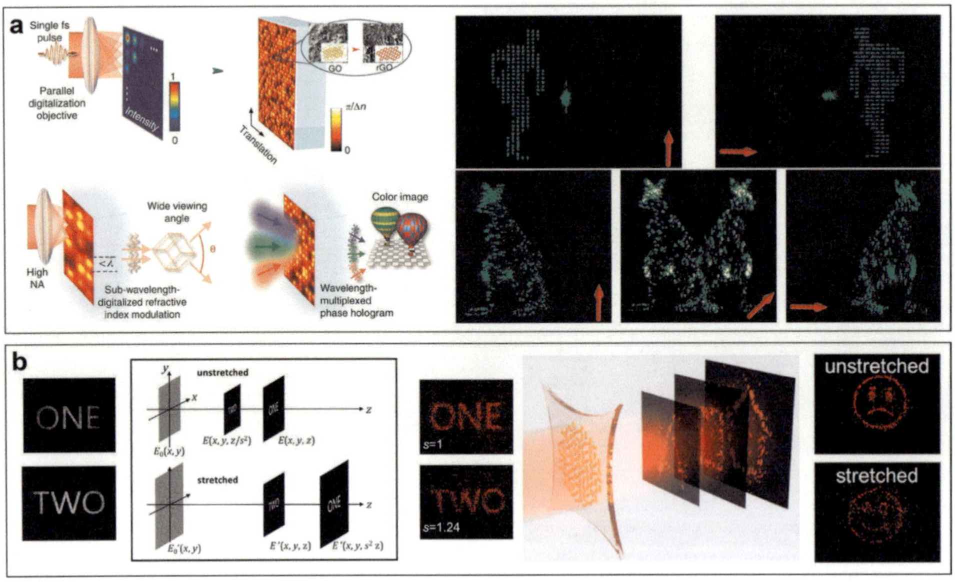

Figure 4.
Methods for rewritable and stretchable meta-holography. (a) A tunable metasurface hologram based on functional graphene oxide materials, enabling 3D holography with wide viewing angles. The hologram is rewritable through femtosecond laser pulse irradiation. (b) Reconfigurable phase-only computer-generated metasurface holograms featuring up to three image planes in the visible spectrum, fabricated using gold nanorods on a stretchable polydimethylsiloxane (PDMS) substrate. Figure reproduced from: (a) Ref. [73], Springer Nature. (b) Ref. [89], American Chemical Society.

images using reduced graphene oxide (**Figure 4a**) [73]. They discovered that the femtosecond (fs) laser intensity at the focal spot directly influenced the refractive index modulation in the reduced graphene oxide. The athermal photoreduction process was confined to a diffraction-limited region, as the fs laser pulse prevented undesired accumulative heating. Furthermore, by precisely controlling laser irradiance, the reversible reduction and oxidation of graphene oxides enabled the creation of updatable reduced graphene oxide holograms.

In 2017, Malek et al. demonstrated another approach to tunable meta-holography [89], developing a reconfigurable meta-hologram by placing gold nanorods on a stretchable polydimethylsiloxane (PDMS) substrate. Stretching the substrate altered the spectral response of the metasurface [87, 88], the focal length of the metalens [90], and the distance of the displayed image from the meta-hologram panel [89]. By encoding multiple reconstructed images at varying distances onto a single metasurface, the researchers enabled dynamic switching of holographic images on the same observation plane through mechanical stretching of the metasurface device (**Figure 4b**).

3. Multiplexed meta-holography

Light possesses multiple fundamental properties that serve as distinct dimensions, including propagation direction, wavelength (frequency), polarization, and OAM. Multiplexing technologies play a crucial role in optical communication and computing, significantly expanding bandwidth and enhancing processing speed. Similarly, multiplexing techniques can be applied in meta-holography to achieve dynamic holographic displays.

3.1 Wavelength-multiplexed meta-holography

Traditional optical holography typically employs a single-wavelength laser due to interference constraints. In 1968, Dr. Stephen A. Benton introduced the rainbow hologram, which utilized a slit to eliminate vertical parallax and minimize spectral blur. The principle of rainbow illumination has also been applied to volumetric imaging using digital dynamic holographic devices, such as spatial light modulators (SLMs), to expand the vertical and longitudinal viewing zones. However, this approach is generally unsuitable for full-color display because computer-generated holography (CGH) devices exhibit wavelength-dependent properties, complicating color image reconstruction. Techniques such as frequency filtering [94] and phase shift [95] have been developed to achieve full-color holography with rainbow illumination.

Achieving a broadband display is crucial for practical holographic applications. Numerous innovative strategies have been developed to realize full-color meta-holographic displays by leveraging the extensive design flexibility of subwavelength structures [96–98] and metasurface elements [99–101].

In 2015, Huang et al. developed a phase-modulated multicolor meta-hologram composed of a two-dimensional pixel array, with each pixel containing four subpixels (two for blue, one for red, and one for green) (**Figure 5a**) [96]. Each subpixel was

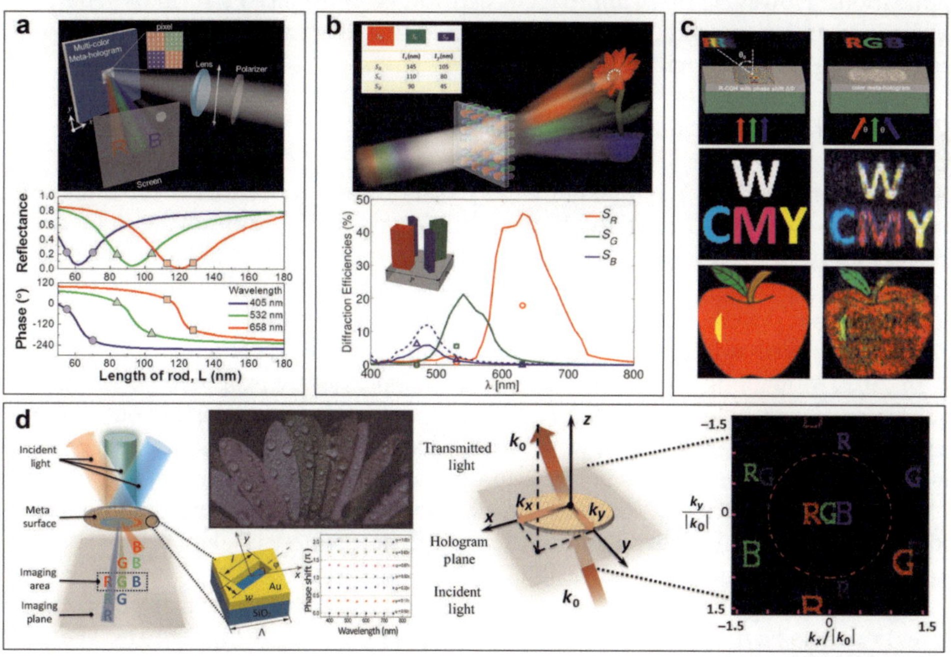

Figure 5.
Wavelength-multiplexed meta-holography. (a) Schematic of a multicolor meta-hologram design under linearly polarized illumination. (b) Diagram of visible-frequency silicon metasurfaces composed of three types of nanoblocks multiplexed within a subwavelength unit. (c) Schematic of full-color plasmonic metasurface holograms incorporating both amplitude and phase modulation for the reconstruction of 2D and 3D color holographic images. (d) Schematic of the off-axis illumination technique used in full-color meta-holography. The top-right image displays experimental results based on plasmonic nanoslit antennas. Figure reproduced from: (a–c) Refs. [96, 97, 100], American Chemical Society. (d) Ref. [99], American Association for the Advancement of Science.

constructed using a 4×4 aluminum nanorod pattern, and different colors were obtained by varying the geometrical dimensions of the nanorods. This approach linked subwavelength geometry with spectral response, avoiding additional fabrication complexities but restricting the phase difference range from 0 to π.

Wang et al. later designed a full phase-controlled multiwavelength meta-hologram using similar principles (**Figure 5b**) [97]. By adjusting nanoblock dimensions and orientations, they modulated response spectra and achieved arbitrary phase delays. As with other diffractive elements, the reconstructed image sizes and diffraction angles were wavelength-dependent.

In 2016, Wan et al. introduced a wavelength-multiplexed full-color meta-hologram based on phase shift encoding (**Figure 5c**) [100]. Their approach incorporated an additional phase shift $\Delta\phi = -k_\lambda \sin(\theta_\lambda)\Delta x$ along the x-axis, which was encoded into the metasurface and correlated with the tilted angle of incident light with wavelength λ. This method effectively eliminated color dispersion without requiring complex geometrical parameters or specialized nanostructured materials. Li et al. further advanced the field by designing a seven-wavelength meta-hologram with a 1.39× larger color gamut than conventional RGB-based designs (**Figure 5d**). Additionally, Zhang et al. combined off-axis illumination with polarization-dependent geometric phase modulation to create a full-color meta-hologram capable of functioning in both transmission and reflection modes [101].

Indeed, most studies on colorful meta-holography have primarily focused on full-color display, data storage, and information encryption rather than dynamic display applications. However, these approaches can still be considered a form of generalized dynamic meta-holography, as the reconstructed images can be switched by varying the wavelength of the incident light.

3.2 Polarization-multiplexed meta-holography

As a transverse wave, light exhibits polarization, which represents the direction of oscillation, particularly the orientation of the electric field. Traditional CGH devices are either polarization-insensitive or operate only under specific polarization states. However, meta-holographic elements with anisotropic subwavelength structures exhibit distinct responses depending on the polarization state of the incident light, making them well-suited for polarization-multiplexed holography.

In 2013, Montelongo et al. and Chen et al. independently proposed polarization-multiplexed meta-holograms [102, 103]. Each polarization state carried out independent diffraction patterns, which were encoded and merged into a single metasurface. By varying the incident light's polarization from 0 to 90°, the reconstructed images could be switched, while at 45°, both images were reconstructed with half the intensity. Montelongo et al. utilized amplitude modulation, whereas Chen et al. implemented a four-phase level design.

Beyond linear polarization states, spin angular momentum (circular polarization states) can also be used for meta-holographic encoding [104–106]. In 2015, Wen et al. developed a helicity-multiplexed meta-hologram with two off-axis images that could be switched by altering the helicity of incident light [107]. Arbabi et al. introduced metasurface devices with complete control over polarization and phase, including polarization-switchable phase holograms [108]. Their design could realize arbitrary phase distributions for two orthogonal polarization states by anisotropic dielectric metasurfaces compared with Chen's work. The input ($\boldsymbol{E_{in}}$) and output waves ($\boldsymbol{E_{out}}$) at each pixel could be related by a general Jones matrix, as $\boldsymbol{E_{out}} = \boldsymbol{TE_{in}}$.

$$T = \begin{bmatrix} T_{xx} & T_{xy} \\ T_{yx} & T_{yy} \end{bmatrix} \tag{1}$$

By implementing any unitary and symmetric Jones matrix at each pixel, arbitrary polarization and phase control can be achieved. Mueller et al. extended this method using a combination of geometric and propagation phases to realize two independent arbitrary phase distributions for any pair of orthogonal polarization states (linear, elliptical, or circular) [109]. They demonstrated this by generating two independent images corresponding to left-handed and right-handed circular polarization states (**Figure 6a**). The geometric phase method can also be integrated with other phase modulation techniques. In 2016, Khorasaninejad et al. designed broadband and chiral binary meta-holograms utilizing detour and geometric phases to produce switchable images for left-handed and right-handed circular polarization states (**Figure 6b**) [110]. Operating with high efficiency (up to 75%), this device covered wavelengths from the visible to the near-infrared range.

Most polarization-dependent meta-holograms focus on only two orthogonal states. In 2018, Zhao et al. expanded this approach by integrating twelve polarization manipulation channels into a single birefringent vectorial meta-hologram [113]. In 2019, Arbabi et al. employed birefringent dielectric metasurfaces to create vectorial holograms with arbitrary polarization states. They mapped color information onto Stokes parameters and reconstructed them as monochromatic images using a modified Gerchberg-Saxton algorithm (**Figure 6c**) [111].

Recent advancements in metasurfaces are primarily driven by progress in classical optical theories, such as the generalized Snell's law. Deng et al. explored the

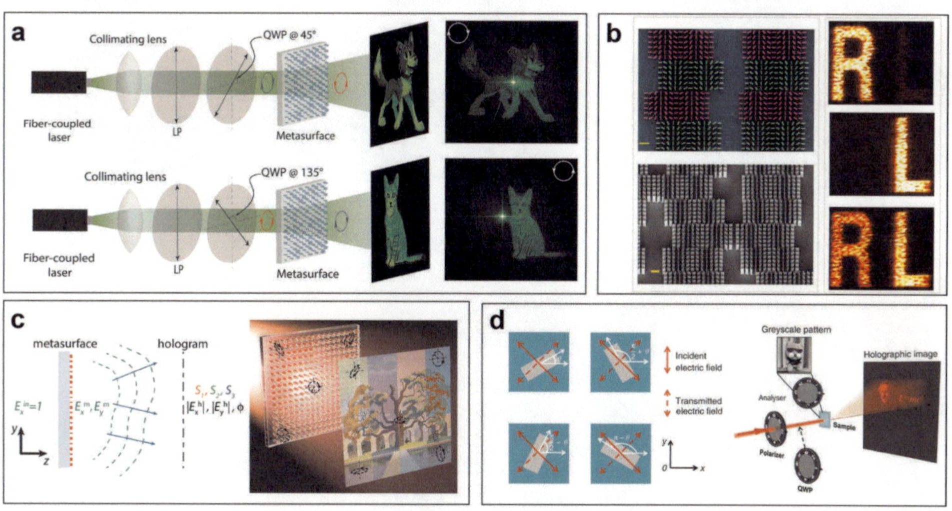

Figure 6.
Polarization-multiplexed meta-holography. (a) Schematic and experimental realization of holographic images of a cartoon dog and cat, achieved using tailored Si nanofins for orthogonal circular polarization multiplexing. (b) Chiral holograms display different images depending on the handedness of the reference beam, enabled through geometric phase encoding. (c) Schematic of a structurally birefringent dielectric metasurface that projects a specific polarization pattern, encoding an RGB image with arbitrary complexity. (d) Schematic of Malus metasurfaces employing a one-to-four mapping and two independent information channels for intensity and phase manipulation. Figure reproduced from: (a) Ref. [109], American Physical Society. (b) Ref. [110], American Association for the Advancement of Science. (c) Ref. [111], American Chemical Society. (d) Ref. [112], Springer Nature.

orientation degeneracy implied by Malus's law and discovered a one-to-M mapping between light irradiance and the Pancharatnam-Berry phase. They proposed a Malus-metasurface-assisted polarization multiplexing method that simultaneously generated a near-field grayscale pattern and an independent far-field holographic image (**Figure 6d**) [112].

These polarization-multiplexed meta-holograms have promising applications in dual-channel displays, anti-counterfeiting, encryption, security, and data storage.

3.3 Angle-multiplexed meta-holography

The angular response characteristics of diffractive optics, including gratings, diffractive optical elements, and metasurfaces, play a crucial role in optical performance. The correlation between input and output angles, governed by physical optics principles, introduces independent channels, providing an additional degree of freedom for multiplexed meta-holography.

In 2017, Kamali et al. designed an angle-multiplexed metasurface using dielectric U-shaped units with high-contrast reflection [114]. The U-shaped nanostructures were optimized to achieve a full 2π phase shift, enabling independent phase control at incident angles of 0 and 30°. This design allowed for two independent channels, each corresponding to a different incident angle, to display separate images (**Figure 7a**).

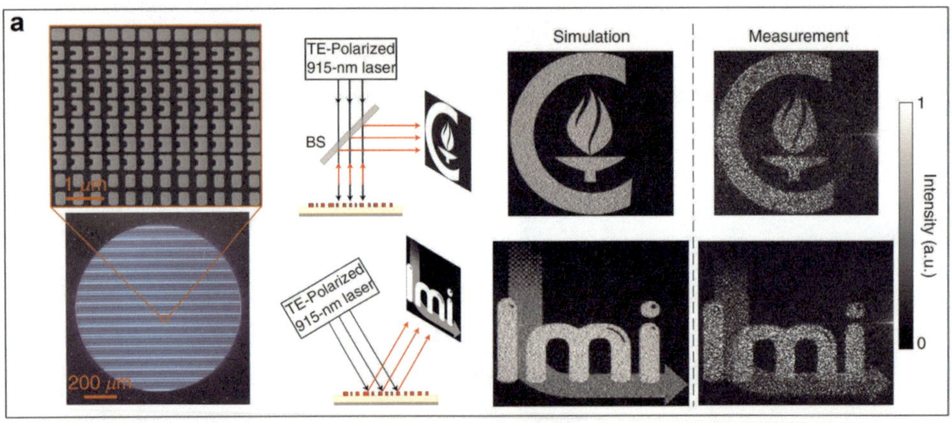

Figure 7.
Angle-multiplexed meta-holography. (a) Angle-multiplexed metasurfaces composed of reflective, high-contrast dielectric U-shaped meta-atoms. (b) Schematic illustration of the coherent pixel polarization meta-holography scheme and the angle-multiplexed metasurface design. Figure reproduced from: (a) Ref. [114], American Physical Society. (b) Ref. [115], WILEY-VCH Verlag GmbH & Co. KGaA, Weinheim.

In 2020, Wang et al. developed an angle-multiplexed multichannel metasurface capable of independently controlling handedness and azimuth, thereby creating four distinct channels (**Figure 7b**) [115]. This design enabled full Poincaré beams, dual-way switching, vectorial print images, and optical information communication. Combining angle multiplexing with other multiplexing strategies further expands the number of available channels. As discussed in Section 3.1, an additional phase shift along a selected axis can be encoded into the metasurface to achieve wavelength-multiplexed meta-holography [99–101]. By varying the incident light's wavelength or angle, phase shifts can be corrected for different wavelengths, allowing images to be displayed at desired positions by simultaneously employing wavelength and angle multiplexing.

Angle multiplexing can also be used to realize other types of dynamic meta-holograms. In 2017, Zhang et al. designed an ultrahigh-capacity meta-hologram by encoding information into nanohole arrays, increasing imaging capacity by a factor of 11.5 compared to conventional metasurfaces [116]. By adjusting the incident angle, compressed image information could be reconstructed, revealing hidden details *via* off-axis illumination. This approach has potential applications in large-capacity optical storage and lithography technologies.

3.4 OAM-multiplexed meta-holography

Orbital angular momentum (OAM) is a fundamental property of photons, characterized by a helical wavefront, a doughnut-shaped intensity distribution, and an unbounded set of orthogonal helical modes, known as topological charge (l). Due to these unique features, OAM plays a crucial role in various applications, including optical communication, stimulated emission depletion (STED) microscopy, and optical tweezers. As a result, OAM can serve as an additional degree of freedom in dynamic meta-holography.

In 2019, Ren et al. proposed an OAM-multiplexed meta-holography design and achieved a dynamic holographic display (**Figure 8a**) [117]. They developed a 10-bit OAM-multiplexed meta-hologram capable of generating 2^{10} distinct OAM-dependent holographic images. In addition to the OAM-multiplexed design, two other types of OAM meta-holograms—OAM-conserving and OAM-selective meta-holograms—were theoretically and experimentally demonstrated. Fang et al. designed similar OAM-preserved, OAM-selective, and OAM-multiplexed holograms based on an SLM platform [121]. In 2019, Ren et al. finished their research work [117], which was based on pure phase modulations, facing the challenge of channel crosstalk and the limitation of the number of channels. Therefore, Ren et al. optimized the method, proposed complex amplitude metasurface-based OAM holography, and exhibited this with 200 separate OAM channels [122]. In this design, the out-of-plane height of the nanopillars provided amplitude control, while their in-plane rotation controlled the phase. The reconstructed images were displayed as two independent, smooth holographic videos at different planes, highlighting the potential for 3D holography.

Also in 2019, Jin et al. proposed a dielectric multimomentum meta-transformer capable of reconstructing different OAM beams with various topological charges into distinct on-axis images (**Figure 8b**) [118]. Although the meta-transformer had a fixed phase profile after fabrication, it synergized the intrinsic properties of light (e.g., OAM and linear momentum) to achieve a dynamic meta-hologram. The total phase profile of the transmitted beam was given by $\psi_T(x_0, y_0) = \psi_{OAM}(x_0, y_0) + \psi_{meta}(x_0, y_0)$, where

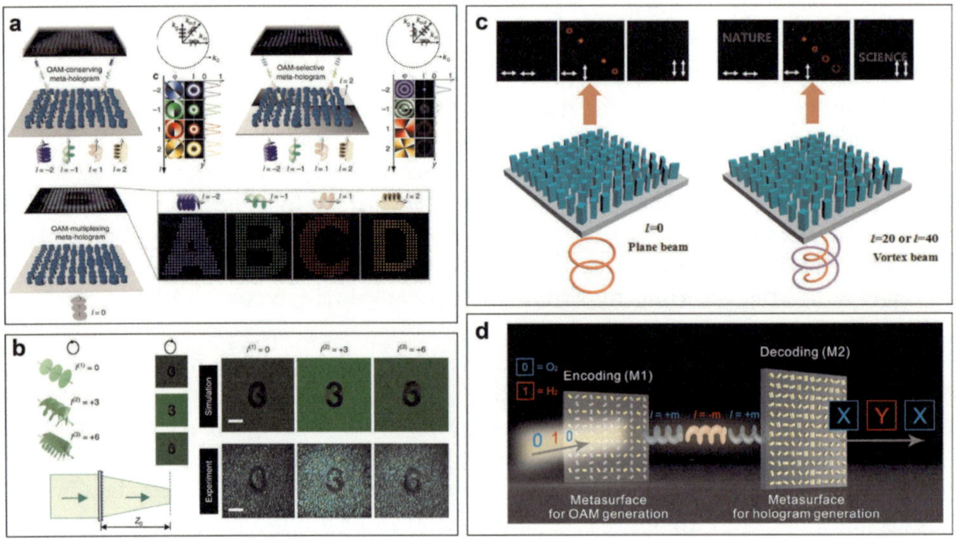

Figure 8.
OAM-multiplexed meta-holography. (a) Schematic of metasurface orbital angular momentum (OAM) holography, utilizing strong OAM selectivity provided by meta-holograms composed of GaN nanopillars with discrete spatial frequency distributions. (b) Schematic of the OAM meta-transformer, designed to operate with three distinct OAM states. (c) Schematic of OAM multiplexing across different polarization channels using a birefringent metasurface for holographic encryption. (d) Schematic of a dynamic metasurface enabling holographic pattern switching via *OAM multiplexing. Figure reproduced from: (a–b) Refs. [117, 118], Springer Nature. (c, d) Refs. [119, 120], American Chemical Society.*

$\psi_{meta}(x_0, y_0)$ was determined by the nanostructure orientation, while $\psi_{OAM}(x_0, y_0)$ could be dynamically modified by switching the topological charge l, which was the phase profile of incident OAM beams. Under red, green, and blue illumination, distinct patterns representing "R," "G," and "B" were reconstructed, demonstrating the feasibility of OAM-based colorful holography.

OAM-multiplexed meta-holography can also be combined with other modulation techniques. In 2020, Zhou et al. proposed a holographic encryption scheme based on OAM-multiplexed meta-holograms across different polarization channels [119]. The selective holographic image could be reconstructed only when the incident light beam had both the correct topological charge l and polarization state (**Figure 8c**). In Section 2.2, we introduced dynamic meta-holography achieved through the metal-to-dielectric phase transition of magnesium (Mg). In 2019, Yu et al. extended this concept by designing a dynamic OAM-multiplexed meta-holography system using two cascaded metasurfaces (**Figure 8d**) [120]. The first metasurface, composed of Mg nanorods, generated switchable vortex beams with different topological charges ($l = m$ and $l = -m$) *via* hydrogenation and dehydrogenation reactions. The second metasurface produced distinct reconstructed images in response to incoming OAM beams with designed topological charges (m and $-m$). This integrated system successfully achieved a dynamic OAM-multiplexed meta-hologram.

3.5 Optical addressed dynamic meta-holography

Beyond conventional pixel-based display screens, several alternative approaches can be used to achieve high-quality 2D displays. The first approach follows a cinematographic method in which different frames from a continuous video are displayed sequentially over time. The second approach involves segmenting a complete image into multiple subgraphs and displaying them in various combinations at different times, as seen in digital tubes used in electronic scoreboards. These common 2D display technologies offer potential inspiration for designing dynamic meta-holograms.

In 2020, Izumi et al. demonstrated a meta-holographic movie based on the first approach [123]. They arrayed 48 meta-hologram channels on a substrate fixed onto a two-axis electric moving stage, which moved at a predefined speed and sequence. By sequentially illuminating each hologram channel, holographic frames were reconstructed on a screen (**Figure 9a**). This method achieved a smooth meta-holographic video with a frame rate of 30 frames per second.

In the same year, Gao et al. proposed an optically addressed dynamic 3D meta-hologram with a large frame capacity and high frame rate using space-channel multiplexing [124]. Their approach divided the meta-hologram samples into numerous spatial channels, each illuminated by a high-speed coded structured light beam. In addition to producing a cinematographic meta-holographic video with 20 frames, they demonstrated an optically addressed space-channel multiplexed meta-hologram with 28-bit encoding, capable of displaying 2^{28} different frames (**Figure 9b**). To achieve high-precision optical addressing, they constructed a dynamic space beam coding

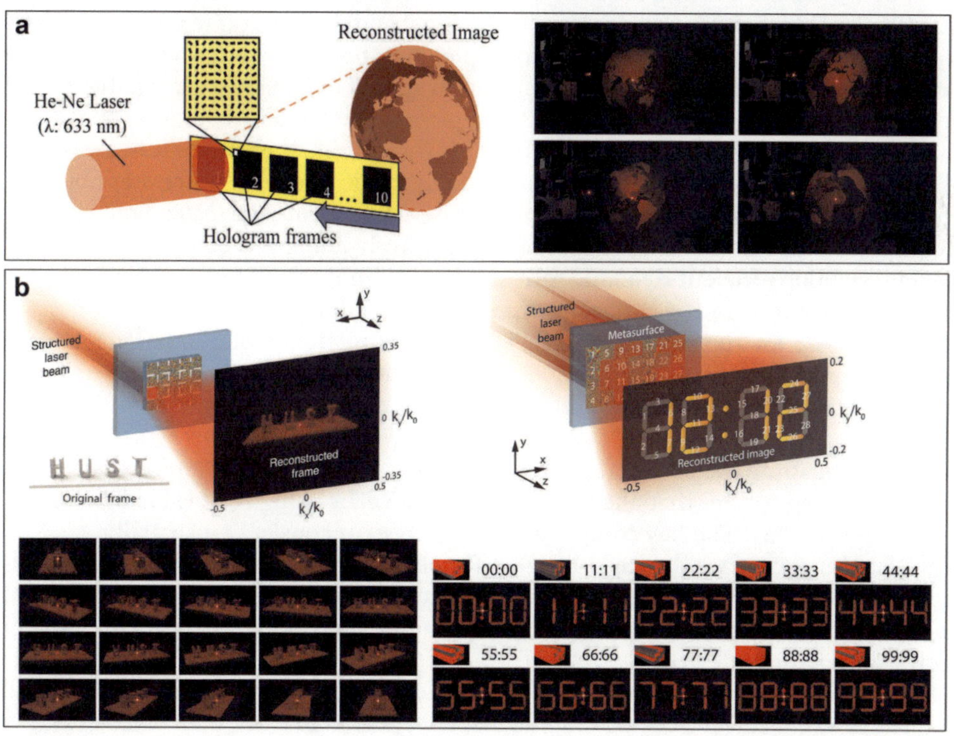

Figure 9.
Optical-addressed space-channel multiplexed meta-holography. (a) Schematic of a cinematography-inspired metasurface holographic movie, where time-lapse images are reconstructed by sequentially illuminating metasurface hologram frames. (b) Schematic of a space-channel multiplexed metasurface, where a structured laser beam selectively activates spatial channels in a predefined sequence, enabling the continuous display of holographic video frames. Figure reproduced from: (a) Ref. [123], Optical Society of America. (b) Ref. [124], American Association for the Advancement of Science.

module comprising a digital micromirror device (DMD), a tube lens, and an objective lens. This system enabled the meta-hologram to operate at an impressive frame rate of 9523 frames per second.

Despite significant progress, dynamic metasurface holography still faces considerable challenges in achieving real-time, highly fluid dynamic display effects required for advanced applications, such as human-computer interaction. The performance of dynamic metasurface holographic displays relies on two critical factors: computational frame rate and display frame rate. On the one hand, the computational frame rate—which refers to the speed of data calculation, processing, and preparation for display—ensures that the system can compute the required content in real time. Current holographic display solutions often rely on performing multiple rapid Fourier transforms (FFTs), typically requiring dedicated computational units such as graphics processing units (GPUs) to achieve high refresh rates. However, the substantial computational demand poses challenges in terms of processing power and energy consumption, limiting the widespread adoption of these solutions. On the other hand, the display frame rate—which determines how quickly a display device refreshes and presents new content—remains a significant hurdle for most metasurface-based dynamic holographic displays. Achieving high display frame rates is essential for delivering a smooth visual experience, yet current strategies struggle to meet this requirement.

In 2024, Liu et al. introduced a dynamic interactive bitwise metasurface holography (Bit-MH) technique with high computational and display frame rates [125]. Based on optically addressed space-channel multiplexing, their approach divided the display functionality of metasurfaces into independent spatial channels, each capable of projecting a sub-holographic pattern. By employing optical addressing for space-channel multiplexing, they converted the on/off states of all channels into a collection of binary values, effectively transforming the dynamic updating process of holography into bitwise operations that govern the respective channels. Instead of relying on frequent FFT calculations, this approach significantly enhanced computational efficiency, enabling efficient dynamic refreshing through mapped bitwise operations.

Benchmark tests of the core algorithm for bitwise dynamic holography were conducted on a low-power Raspberry Pi computing platform, demonstrating a maximum computational frame rate of 800 kHz. Additionally, by utilizing high-speed DMD optical addressing devices, they achieved a maximum display frame rate of 23 kHz.

To demonstrate the concept, the research team constructed an interactive holographic gaming system that allowed users to play Tetris using the visible light spectrum (**Figure 10a**). The key components of the system consist of a spatially segmented metasurface device, a DMD, a Raspberry Pi controller, a gaming controller, and other essential optical components. The proposed design for bitwise dynamic holography enables efficient updates of holographic images and real-time interaction with external input devices.

This efficient and programmable Bit-MH method is expected to serve as a foundation for future metasurface holographic display systems that are both smooth and computationally efficient. Since Bit-MH focuses on manipulating the states of multiple channels, it can be seamlessly integrated with other multichannel multiplexing techniques, further expanding its potential applications.

Information capacity is as crucial as speed and frame rate for achieving real-time, high-quality dynamic displays, as it determines the richness and volume of content that can be transmitted instantly. The technique introduced by Liu et al. provides a

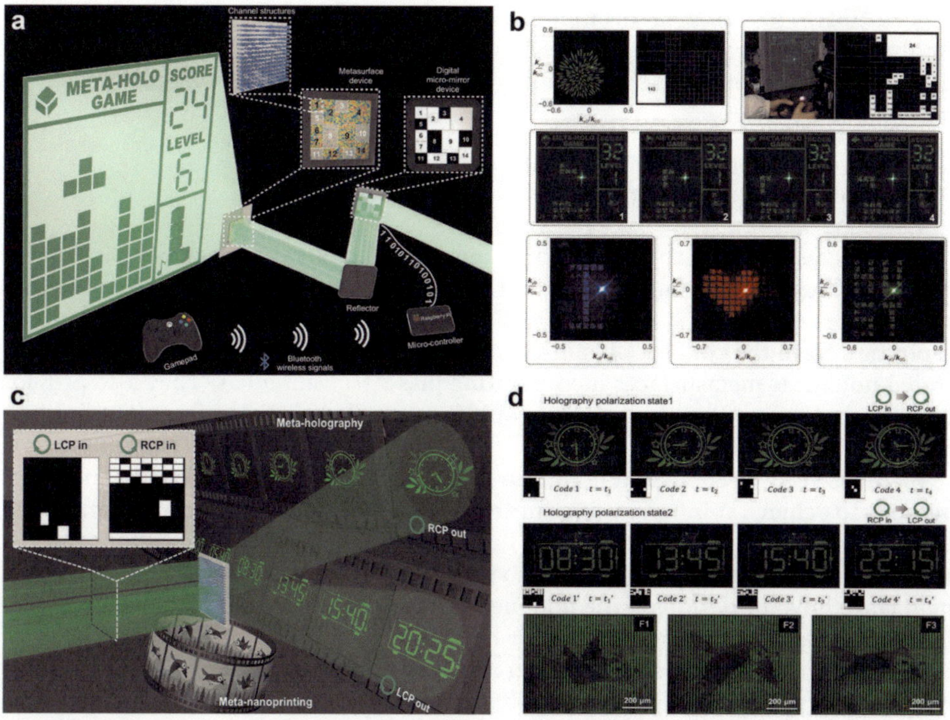

Figure 10.
Optical-addressed dynamic interactive bitwise meta-holography. (a) Schematic diagram of the interactive dynamic metasurface holographic game system, utilizing bitwise meta-holography. (b) Experimental results of reconstructed patterns at wavelengths of 473, 532, and 633 nm, with partial channels activated. (c) Operation schematic of the polarization-multiplexed dynamic display for simultaneous meta-holography and meta-nanoprinting. (d) Experimental results of the polarization-dependent metasurface with different space channels for dynamic meta-holography and meta-nanoprinting. Figure reproduced from: (a, b) Ref. [125], Opto-Electronic Journals Group. (c, d) Ref. [126], WILEY-VCH Verlag GmbH & Co. KGaA, Weinheim.

flexible and robust foundation for extending display functionality in high-speed, high-frame-rate systems. Building upon this foundation, Xu et al. developed a method that combines high-speed, optically addressable dynamic holography with polarization multiplexing [126]. This approach not only achieves an extremely high frame rate but also significantly enhances information capacity by enabling simultaneous, independent displays across multiple polarization states on a single metasurface.

In this design, the metasurface is configured to operate at three independent polarization states, allowing for dynamic holographic display at two orthogonal circular polarization states while simultaneously supporting grayscale meta-nanoprinting at a linear polarization state (**Figure 10c**). By employing polarization multiplexing, the system effectively decouples display functions across distinct spatial channels on a single metasurface. This setup enables independent displays for each polarization state with ultra-high frame rates, reaching 9523 frames per second, thereby expanding the metasurface's information capacity without compromising resolution.

To achieve this, the metasurface incorporates nanoscale structural designs that respond uniquely to each polarization state, modulating the light field and dynamically displaying complex patterns. Experimental results demonstrated the feasibility of this approach by successfully displaying a "clock" and "digital clock" under circular polarization while achieving near-field grayscale imaging under linear polarization

(**Figure 10d**). This polarization-multiplexed strategy not only enhances frame rates but also paves the way for applications in optical encryption, virtual reality, and high-capacity information storage by enabling real-time, high-fidelity dynamic displays on a compact platform.

3.6 Nonlinear wavelength-multiplexed meta-holography

Most metasurface research, including meta-holography, has been conducted within the linear optics regime. In recent years, however, researchers have extended metasurface studies to the nonlinear optics regime [127–129]. In 2015, Li et al. drew inspiration from the concept of spin-rotation coupling and experimentally demonstrated nonlinear metasurfaces incorporating plasmonic antennas. They also provided theoretical evidence that this concept could be extended to dielectric and bulk-like metamaterials [127].

In 2016, Ye et al. combined nonlinear metasurfaces with several multiplexing methods to achieve spin- and wavelength-multiplexed nonlinear meta-holography [130]. Nonlinear meta-holography can be regarded as a specialized multiplexing technique, as these metasurfaces remain static. Split-ring resonators were chosen as the meta-atoms for designing the nonlinear metasurface due to their robust polarization properties in linear optics and their high efficiency in second-harmonic generation. The reported design enabled the creation of multiple independent holographic images using fundamental and second-harmonic waves with different spin states, effectively separating them into distinct channels (**Figure 11a**). Notably, nonlinear meta-holography provides independent and crosstalk-free channels for holographic

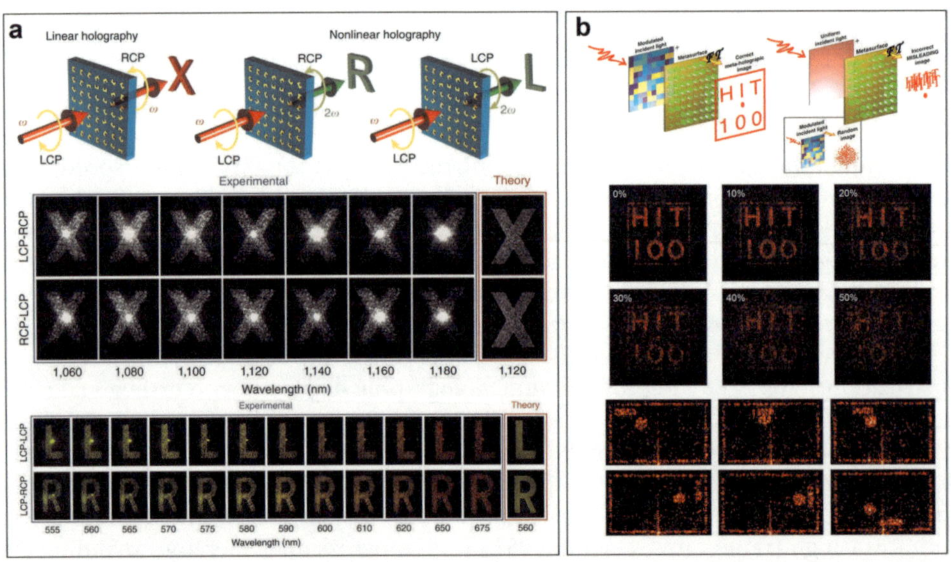

Figure 11.
Nonlinear and reprogrammable meta-holography. (a) Nonlinear holographic metasurfaces: *Linear holography reconstructs the letter "**X**" at infrared wavelengths, while nonlinear holography generates distinct letters encoded in different circular polarization states. (b)* Reprogrammable metasurface: *Schematic of a metasurface capable of displaying arbitrary holographic images for* optical encryption. *The holographic image is revealed when illuminated with a modulated incident beam but remains hidden under an incorrect uniform laser beam. Figure reproduced from: (a, b) Refs. [130, 131], Springer Nature.*

reconstruction, making it well-suited for applications such as multichannel holographic displays, optical data storage, and optical encryption.

3.7 Incident light field multiplexed meta-holography

In the aforementioned studies, the encoded data for holographic reconstruction was generated using metasurface elements, rather than being embedded in the incident light beams. Even in methods that involve incident light modulation (e.g., OAM multiplexing) [117, 118, 122], the OAM beam—defined by a specific topological charge—was used solely to select and activate the desired channel, without directly carrying information for holographic reconstruction.

In 2020, Qu et al. introduced a novel approach to incident light field-multiplexed meta-holography, referred to as the "reprogrammable meta-hologram" [131]. Their method divided holographic information into two matrices: One was conventionally encoded into the metasurface, while the other was encoded into the incident light beam. By leveraging the phase matrix of the incident light as an additional degree of freedom, this approach enabled the dynamic generation of a wide range of holographic images and videos. As a result, a fixed metasurface, when used in conjunction with a modifiable incident light beam, could achieve extensive reconfigurability in meta-holographic displays (**Figure 11b**).

3.8 Multi-dimensional multiplexed meta-holography

To enhance the number of frames and encoding flexibility in practical holographic applications, various techniques can be employed to design multiple meta-hologram elements simultaneously. As discussed earlier, Li et al. demonstrated a colorful meta-hologram by combining wavelength and angle multiplexing, while Yu et al. proposed a dynamic meta-holography design based on two metasurface elements—one achieved through a chemical reaction method and the other utilizing OAM multiplexing.

In 2018, Jin et al. introduced (2^6–1) spin- and wavelength-encoded holograms, where they simultaneously manipulated six bases of incident photons to reconstruct 6-bit spin- and wavelength-dependent multicolor images (**Figure 12a**) [132]. Furthermore, in 2020, Deng et al. proposed a multi-degree-of-freedom metasurface technique by integrating the geometric Pancharatnam-Berry phase with the detour phase. This approach enabled simultaneous control of both the wavefront's complex amplitude and polarization, exhibiting complete independence from both wavelength and incident angle. As a result, they successfully demonstrated a versatile, full-color, complex amplitude vectorial meta-hologram [134].

In 2024, Liang et al. proposed a tri-channel metasurface that can further extend the meta-imaging ranges, with three independent images located at the interface, Fresnel, and Fourier domains, respectively. Through this approach, they encoded three distinct images into a single metasurface, including a color nanoprinting image, a Fresnel holographic image, and a Fourier holographic image (**Figure 12b**) [133]. With its compact design, easiness of fabrication, minimal crosstalk, and substantial storage density, the tri-channel metasurface presents exciting prospects in various fields, spanning from image display and data storage to information encryption and anti-counterfeiting measures.

Beyond dynamic holographic displays, researchers have also explored the integration of meta-holography with other imaging methods. In 2020, Li et al. experimentally

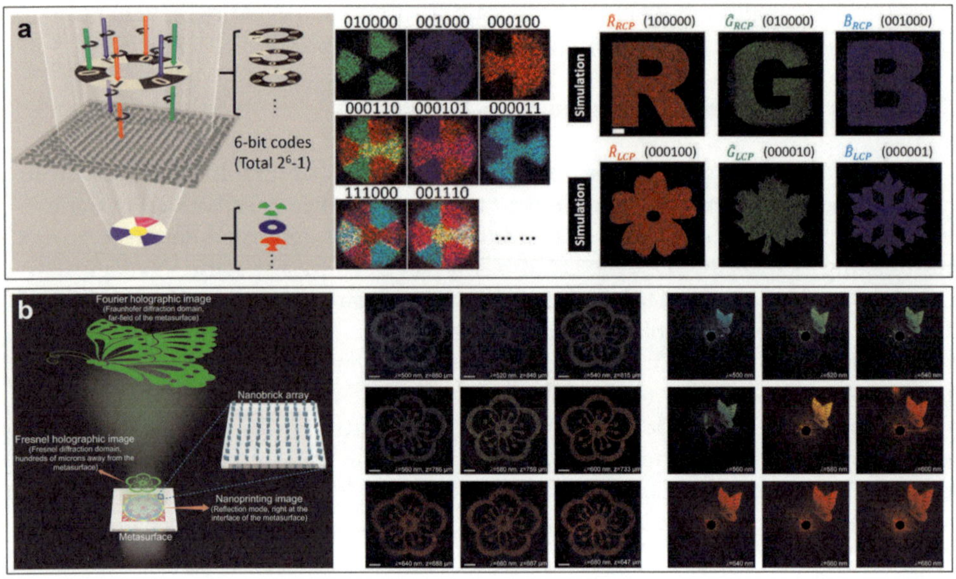

Figure 12.
Multi-dimensional multiplexed meta-holography. (a) Schematic of (2^6–1) spin- and wavelength-encoded holograms, where six bases of incident photons are simultaneously manipulated to reconstruct 6-bit multicolor images dependent on both wavelength and spin. (b) A tri-channel metasurface with three independent images located at the interface, Fresnel, and Fourier domains, respectively. A structural-color nanoprinting image is decoded right at the interface of the metasurface; a Fresnel holographic image and another Fourier holographic image are present at the Fresnel and Fourier (far-field) domains, respectively. Figure reproduced from: (a) Ref. [132], Springer Nature. (b) Ref. [133], Walter Gruyter GmbH.

demonstrated a three-channel metasurface capable of simultaneously recording a continuous grayscale nanoprinting image in the near-field while projecting two independent holographic images in the far field [135]. In the same year, Dai et al. introduced a "three-in-one" metasurface device, which integrated a structural-color nanoprint element, a polarization-controlled grayscale metasurface image display, and a phase-modulated meta-hologram into a single unit [136]. These pioneering studies have inspired further exploration of novel meta-holographic techniques, broadening the potential applications of metasurfaces.

4. Other applications of the meta-holography

Metasurface holography offers a versatile and programmable approach to manipulating light fields at the nanoscale, enabling advancements beyond displays in fields such as optical communications, optical imaging and sensing, and advancing manufacturing. This section provides a brief overview of some representative applications of meta-holography in other domains.

4.1 Imaging and sensing

Spectrum and polarization are key attributes that reveal vital information about light propagation. Spectral imaging, for instance, uncovers material composition, while polarization imaging informs surface texture and optical characteristics. As such, multispectral and polarization imaging are crucial for fields like archaeology,

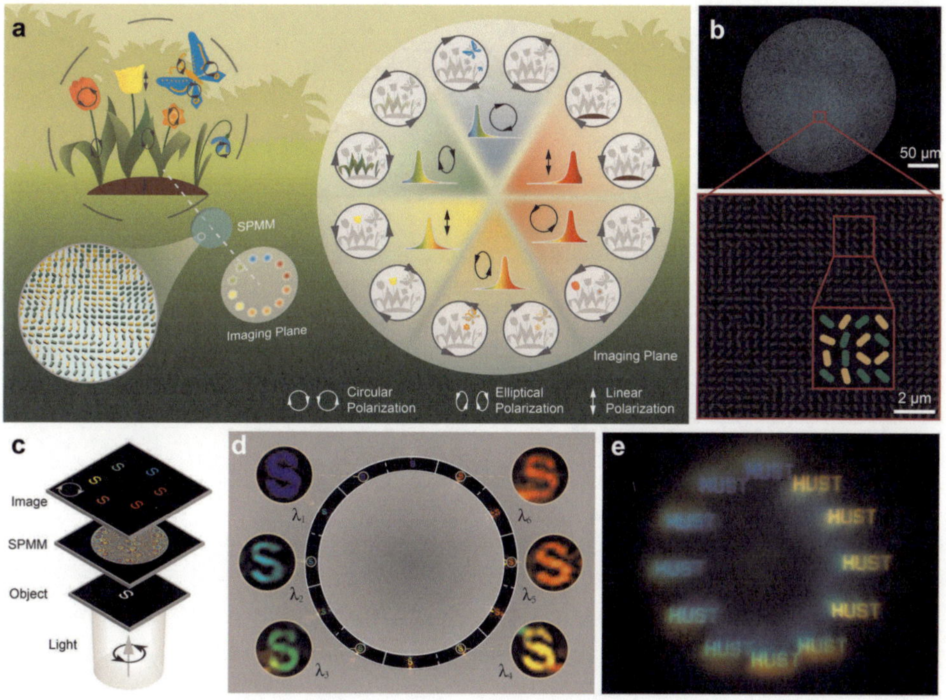

Figure 13.
Multi-dimensional imaging based on meta-holography. (a) The design concept of the SPMM device. The SPMM generates twelve focal points or images corresponding to six spectral bands and two circular polarization states, thereby permitting the reconstruction of features lost in traditional imaging systems. (b) The optical image and SEM image of SPMM. (c) The schematic illustration of the experimental imaging process. (d) Imaging results of the SPMM for an incident laser beam with six wavelengths. (e) Imaging results for the phrase "HUST" using an incoherent flashlight from a mobile phone as the light source. Figure reproduced from: Ref. [137], Opto-Electronic Journals Group.

biology, remote sensing, and astronomy. Conventional techniques using filters and analyzers have demonstrated advantages over basic amplitude imaging, yet they often require multiple exposures and bulky configurations, limiting integration into compact systems.

To address this, Hui Gao et al. introduced the spectrum- and polarization-ellipticity-resolved multifocal metalens (SPMM), enabling single-shot, spectral- and polarization-resolved imaging without moving parts (**Figure 13a**) [137]. By generating twelve distinct spectral- and polarization-dependent images in one exposure, the SPMM efficiently captures optical data. Its design uses space-variant anisotropic subwavelength structures to achieve polarization conversion and spectral differentiation while maintaining traditional focusing capabilities and enhancing imaging performance through a shared-aperture design with a high numerical aperture. Tests with both coherent and incoherent light sources confirm the SPMM's broad applicability, establishing a pathway for compact, multifunctional optical systems poised to advance diverse applications.

4.2 Optical communications

Vortex laser beams with OAM have gained interest due to their distinct phase and intensity profiles, characterized by a hollow light field and an azimuthal phase term

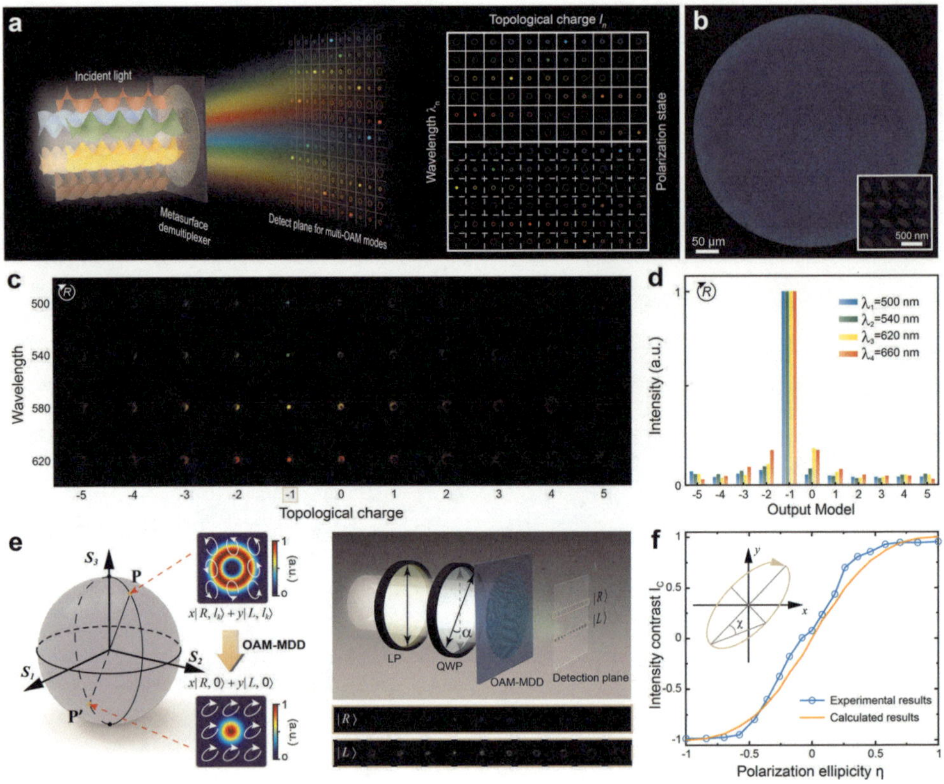

Figure 14.
Optical communications element based on meta-holography. (a) Schematic of the OAM-MDD. Incident light containing multiple OAM beams illuminates the metasurface, generating a multifocal diffraction pattern at the detection plane. (b) Optical image of the OAM-MDD, with an inset SEM image showing spatially variant rectangular meta-atoms. (c, d) Performance characterization of the OAM-MDD, including a histogram of central intensities for an input laser beam containing four wavelengths and an OAM mode of −1. (e) Schematic of the polarization ellipticity demultiplexing process. A hybrid-order Poincaré sphere illustrates the input OAM with a polarization ellipse and its corresponding antithesis point after passing through the OAM-MDD. (f) Polarization state measurements at 540 nm as the incident polarization gradually varies from right circular polarization to left circular polarization. Figure reproduced from: Ref. [138], WILEY-VCH Verlag GmbH & Co. KGaA, Weinheim.

$\exp.(il\varphi)$, where l represents the topological charge and φ the azimuthal angle. Theoretically, OAM beams can support an infinite number of orthogonal states, making them valuable for encoding numerous data bits or creating independent communication channels. However, current OAM-based methods in free-space optical communication, OAM shift keying and OAM division multiplexing, are limited to specific polarization states or wavelengths, restricting compatibility with other multiplexing techniques.

To address these constraints, Hui Gao et al. introduced a single-layer dielectric metasurface based on meta-holography for multi-dimensional demultiplexing and decoding, which eliminates the need for bulk optical elements [138]. This OAM metasurface-based demultiplexer and decoder (OAM-MDD) separates light by wavelength, spin angular momentum (SAM), and OAM, creating distinct foci on the focal plane for each unique mode (**Figure 14**). The experiment validated 132 independent channels based on 6 wavelengths, 11 OAM values, and 2 SAM states and demonstrated elliptic polarization decoding with a 4-bit coded transmission. This compact platform enables multi-dimensional demultiplexing across various wavelengths, with

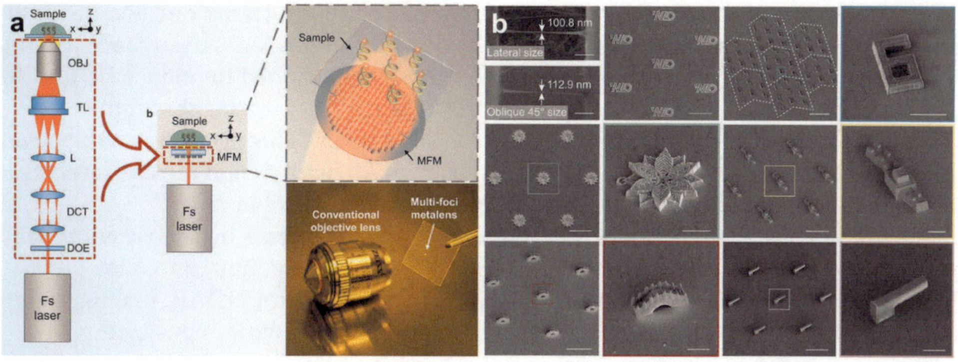

Figure 15.
Advanced laser manufacturing based on meta-holography. (a) Schematic diagram of multi-focus TPP system based on the MFM. (b) SEM characterization results of micro- and nanostructures fabricated based on a multifocal TPP system. Figure reproduced from: (a, b) Ref. [139], WILEY-VCH Verlag GmbH & Co. KGaA, Weinheim.

applications in optical communications, data storage, encryption, and quantum information science.

4.3 Advanced laser manufacturing

Two-photon polymerization (TPP) has emerged as a powerful technique for achieving high-resolution 3D nanolithography, essential for fields like aerospace and biomedical engineering. However, traditional TPP systems are often hindered by bulky, complex optical setups that include multiple components, such as beam splitters and beam expanders, which limit their integration into compact, lightweight, and cost-effective designs.

To address these limitations, Wang et al. recently developed a holographic multi-focus metalens (MFM) approach that simplifies TPP by consolidating multiple optical functions into a single metasurface element [139]. This MFM, constructed with SiNx nanopillars, produces seven uniform, precisely controlled foci that enable efficient, parallel micro- and nanofabrication. Each focus maintains consistent optical characteristics, supporting the reliable creation of sub-diffraction-limited features critical for high-precision 3D structures (**Figure 15a**). By replacing multiple optical elements with a single MFM, this approach enhances the compactness and functionality of TPP systems, offering a promising path toward accessible, high-precision 3D printing and potentially accelerating progress across various scientific and industrial fields (**Figure 15b**).

5. Conclusions

In summary, this chapter has examined recent advancements in dynamic meta-holography, highlighting the advantages of metasurfaces over conventional CGH devices. These advantages include a wide field of view, high resolution, and low noise, as well as multiple degrees of freedom in modulating light's amplitude, phase, and polarization. Such capabilities enable the design of highly versatile dynamic holograms. This review has discussed various research directions in dynamic meta-holography, including tunable, multiplexed, and multi-dimensional metasurfaces.

Achieving a seamless holographic display requires a high frame rate and an abundant supply of high-quality frames. While significant progress has been made, current techniques still fall short of the holographic displays envisioned in science fiction. An ideal solution would involve the rapid control of wave interactions through metasurface nanostructures, akin to LED or LCD screens. This necessitates the development of high-performance metasurface-based SLMs with a high refresh rate, efficient modulation, and broad spectral response across the visible range.

Recent studies have demonstrated promising advancements in electrically tunable metasurface-based SLMs. However, these designs are largely limited to one-dimensional applications such as beam steering, focusing, and LiDAR, making them unsuitable for full holographic displays. The fabrication of two-dimensional metasurface-based SLMs in the visible range remains a challenge due to limitations in current nanofabrication techniques. As an alternative, approaches such as OAM multiplexing, space-channel multiplexing, and reprogrammable metasurfaces have emerged, offering potential solutions for dynamic meta-holography in specialized applications. These methods demonstrate significant advantages in terms of frame number and frame rate performance.

With continued advancements in nanofabrication technologies and computational design methodologies, the realization of ideal dynamic meta-holography solutions is expected in the near future.

Beyond its established role in holographic displays, metasurface holography is poised to revolutionize a broad range of fields through its unparalleled ability to manipulate light at the nanoscale. By offering precise control over multiple optical properties, metasurfaces introduce new dimensions of flexibility and efficiency, pushing the boundaries of conventional optical systems. Their compactness, scalability, and high integration capability make them ideally suited for applications requiring miniaturization and versatility, enabling seamless integration into increasingly complex technological ecosystems.

As meta-holography continues to evolve, its applications are extending into domains where traditional optics face limitations. This technology is driving the development of faster, more secure, and highly customizable optical systems, paving the way for breakthroughs in communications, imaging, sensing, and advanced manufacturing. The continuous progress in metasurface holography points toward a future where dynamic, high-resolution, and multifunctional optical platforms become standard, shaping advancements in both scientific research and industry. With its adaptability and transformative potential, metasurface holography is not only reshaping modern optics but also unlocking new frontiers for technological innovation across a wide array of disciplines.

Acknowledgements

The author is grateful for financial support from the National Natural Science Foundation of China (Grant No.62205117), the Young Elite Scientists Sponsorship Program by CAST (Grant No. 2022QNRC001), the West Light Foundation of the Chinese Academy of Sciences (Grant No. xbzg-zdsys-202206), the Knowledge Innovation Program of Wuhan-Shuguang, and the Hubei Provincial Natural Science Foundation of China (Grant No. 2022CFB792). The author also would like to express his gratitude to his students for their help in revising the manuscript of this chapter (in alphabetical order): Xuhao Fan, Xinger Wang, Yuncheng Liu, and Yu Wang.

Conflict of interest

The authors declare no conflict of interest.

Author details

Hui Gao
Wuhan National Laboratory for Optoelectronics, School of Optical and Electronic Information, Huazhong University of Science and Technology, Wuhan, China

*Address all correspondence to: gaohui_wnlo@hust.edu.cn

References

[1] Matharu AS, Jeeva S, Ramanujam PS. Liquid crystals for holographic optical data storage. Chemical Society Reviews. 2007;**36**:1868. DOI: 10.1039/B706242G

[2] Hvilsted S, Sánchez C, Alcalá R. The volume holographic optical storage potential in azobenzene containing polymers. Journal of Materials Chemistry. 2009;**19**:6641. DOI: 10.1039/B900930M

[3] Chen L, Zhao D. Optical color image encryption by wavelength multiplexing and lensless Fresnel transform holograms. Optics Express. 2006;**14**:8552. DOI: 10.1364/OE.14.008552

[4] Georgi P, Wei Q, Sain B, Schlickriede C, Wang Y, Huang L, et al. Optical secret sharing with cascaded metasurface holography. Science Advances. 2021;7:eabf9718. DOI: 10.1126/sciadv.abf9718

[5] Rivenson Y, Zhang Y, Günaydın H, Teng D, Ozcan A. Phase recovery and holographic image reconstruction using deep learning in neural networks. Light: Science & Applications. 2017;7:17141-17141. DOI: 10.1038/lsa.2017.141

[6] Tanriover I, Dereshgi SA, Aydin K. Metasurface enabled broadband all optical edge detection in visible frequencies. Nature Communications. 2023;**14**:6484. DOI: 10.1038/s41467-023-42271-w

[7] Leach J, Sinclair G, Jordan P, Courtial J, Padgett MJ, Cooper J, et al. 3D manipulation of particles into crystal structures using holographic optical tweezers. Optics Express. 2004;**12**:220. DOI: 10.1364/OPEX.12.000220

[8] Curtis JE, Koss BA, Grier DG. Dynamic holographic optical tweezers. Optics Communications. 2002;**207**:169-175. DOI: 10.1016/S0030-4018(02)01524-9

[9] Gabor D. A new microscopic principle. Nature. 1948;**161**:777-778. DOI: 10.1038/161777a0

[10] Brown BR, Lohmann AW. Complex spatial filtering with binary masks. Applied Optics. 1966;**5**:967. DOI: 10.1364/AO.5.000967

[11] Mok F, Diep J, Liu HK, Psaltis D. Real-time computer-generated hologram by means of liquid-crystal television spatial light modulator. Optics Letters. 1986;**11**:748. DOI: 10.1364/OL.11.000748

[12] Hahn J, Kim H, Lim Y, Park G, Lee B. Wide viewing angle dynamic holographic stereogram with a curved array of spatial light modulators. Optics Express. 2008;**16**:12372. DOI: 10.1364/OE.16.012372

[13] Huang L, Zhang S, Zentgraf T. Metasurface holography: From fundamentals to applications. Nano. 2018;7:1169-1190. DOI: 10.1515/nanoph-2017-0118

[14] Zhao R, Huang L, Wang Y. Recent advances in multi-dimensional metasurfaces holographic technologies. PhotoniX. 2020;**1**:20. DOI: 10.1186/s43074-020-00020-y

[15] Dorrah AH, Capasso F. Tunable structured light with flat optics. Science. 2022;**376**:eabi6860. DOI: 10.1126/science.abi6860

[16] Zhang Q, He Z, Xie Z, Tan Q, Sheng Y, Jin G, et al. Diffractive optical elements 75 years on: From micro-optics to metasurfaces. Photonics Insights. 2023;**2**:R09. DOI: 10.3788/PI.2023.R09

[17] Brongersma ML, Pala RA, Altug H, Capasso F, Chen WT, Majumdar A, et al. The second optical metasurface revolution: Moving from science to technology. Nature Reviews Electrical Engineering. 2025;**2**:125-143. DOI: 10.1038/s44287-024-00136-4

[18] Cai T, Wang GM, Zhang XF, Liang JG, Zhuang YQ, Liu D, et al. Ultra-thin polarization beam splitter using 2-D Transmissive phase gradient Metasurface. IEEE Transactions on Antennas and Propagation. 2015;**63**: 5629-5636. DOI: 10.1109/TAP.2015.2496115

[19] Khorasaninejad M, Zhu W, Crozier KB. Efficient polarization beam splitter pixels based on a dielectric metasurface. Optica. 2015;**2**:376. DOI: 10.1364/OPTICA.2.000376

[20] Zhao Z, Pu M, Gao H, Jin J, Li X, Ma X, et al. Multispectral optical metasurfaces enabled by achromatic phase transition. Scientific Reports. 2015;**5**:15781. DOI: 10.1038/srep15781

[21] Gao H, Pu M, Li X, Ma X, Zhao Z, Guo Y, et al. Super-resolution imaging with a Bessel lens realized by a geometric metasurface. Optics Express. 2017;**25**: 13933. DOI: 10.1364/OE.25.013933

[22] Khorasaninejad M, Chen WT, Devlin RC, Oh J, Zhu AY, Capasso F. Metalenses at visible wavelengths: Diffraction-limited focusing and subwavelength resolution imaging. Science. 2016;**352**:1190-1194. DOI: 10.1126/science.aaf6644

[23] Khorasaninejad M, Capasso F. Metalenses: Versatile multifunctional photonic components. Science. 2017;**358**: eaam8100. DOI: 10.1126/science.aam8100

[24] Shrestha S, Overvig AC, Lu M, Stein A, Yu N. Broadband achromatic dielectric metalenses. Light: Science & Applications. 2018;**7**:85. DOI: 10.1038/s41377-018-0078-x

[25] Wang S, Wu PC, Su VC, Lai YC, Chen MK, Kuo HY, et al. A broadband achromatic metalens in the visible. Nature Nanotechnology. 2018;**13**: 227-232. DOI: 10.1038/s41565-017-0052-4

[26] Pu M, Li X, Ma X, Wang Y, Zhao Z, Wang C, et al. Catenary optics for achromatic generation of perfect optical angular momentum. Science Advances. 2015;**1**:e1500396. DOI: 10.1126/sciadv.1500396

[27] Gao H, Li Y, Chen L, Jin J, Pu M, Li X, et al. Quasi-Talbot effect of orbital angular momentum beams for generation of optical vortex arrays by multiplexing metasurface design. Nanoscale. 2018;**10**:666-671. DOI: 10.1039/C7NR07873K

[28] Jin J, Luo J, Zhang X, Gao H, Li X, Pu M, et al. Generation and detection of orbital angular momentum via metasurface. Scientific Reports. 2016;**6**: 24286. DOI: 10.1038/srep24286

[29] Yang K, Pu M, Li X, Ma X, Luo J, Gao H, et al. Wavelength-selective orbital angular momentum generation based on a plasmonic metasurface. Nanoscale. 2016;**8**:12267-12271. DOI: 10.1039/C5NR09209D

[30] Karimi E, Schulz SA, De Leon I, Qassim H, Upham J, Boyd RW. Generating optical orbital angular momentum at visible wavelengths using a plasmonic metasurface. Light: Science & Applications. 2014;**3**:e167-e167. DOI: 10.1038/lsa.2014.48

[31] Li G, Kang M, Chen S, Zhang S, Pun EYB, Cheah KW, et al. Spin-enabled Plasmonic Metasurfaces for

manipulating orbital angular momentum of light. Nano Letters. 2013;**13**: 4148-4151. DOI: 10.1021/nl401734r

[32] Zhu X, Yan W, Levy U, Mortensen NA, Kristensen A. Resonant laser printing of structural colors on high-index dielectric metasurfaces. Science Advances. 2017;**3**:e1602487. DOI: 10.1126/sciadv.1602487

[33] Sun S, Zhou Z, Zhang C, Gao Y, Duan Z, Xiao S, et al. All-dielectric full-color printing with TiO_2 Metasurfaces. ACS Nano. 2017;**11**:4445-4452. DOI: 10.1021/acsnano.7b00415

[34] Huo P, Song M, Zhu W, Zhang C, Chen L, Lezec HJ, et al. Photorealistic full-color nanopainting enabled by a low-loss metasurface. Optica. 2020;**7**: 1171. DOI: 10.1364/OPTICA. 403092

[35] Cheng F, Gao J, Luk TS, Yang X. Structural color printing based on plasmonic metasurfaces of perfect light absorption. Scientific Reports. 2015;**5**:11045. DOI: 10.1038/ srep11045

[36] Proust J, Bedu F, Gallas B, Ozerov I, Bonod N. All-dielectric colored metasurfaces with silicon Mie resonators. ACS Nano. 2016;**10**: 7761-7767. DOI: 10.1021/ acsnano.6b03207

[37] Lee G, Sung J, Lee B. Recent advances in metasurface hologram technologies (invited paper). ETRI Journal. 2019;**41**:10-22. DOI: 10.4218/ etrij.2018-0532

[38] Zheng G, Mühlenbernd H, Kenney M, Li G, Zentgraf T, Zhang S. Metasurface holograms reaching 80% efficiency. Nature Nanotechnology. 2015;**10**:308-312. DOI: 10.1038/ nnano.2015.2

[39] Huang L, Chen X, Mühlenbernd H, Zhang H, Chen S, Bai B, et al. Three-dimensional optical holography using a plasmonic metasurface. Nature Communications. 2013;**4**:2808. DOI: 10.1038/ncomms3808

[40] Wang L, Kruk S, Tang H, Li T, Kravchenko I, Neshev DN, et al. Grayscale transparent metasurface holograms. Optica. 2016;**3**:1504. DOI: 10.1364/OPTICA.3.001504

[41] Zhang X, Jin J, Wang Y, Pu M, Li X, Zhao Z, et al. Metasurface-based broadband hologram with high tolerance to fabrication errors. Scientific Reports. 2016;**6**:19856. DOI: 10.1038/srep19856

[42] Devlin RC, Khorasaninejad M, Chen WT, Oh J, Capasso F. Broadband high-efficiency dielectric metasurfaces for the visible spectrum. National Academy of Sciences of the United States of America. 2016;**113**: 10473-10478. DOI: 10.1073/ pnas.1611740113

[43] Huang K, Liu H, Garcia-Vidal FJ, Hong M, Luk'yanchuk B, Teng J, et al. Ultrahigh-capacity non-periodic photon sieves operating in visible light. Nature Communications. 2015;**6**:7059. DOI: 10.1038/ncomms8059

[44] Butt H, Montelongo Y, Butler T, Rajesekharan R, Dai Q, Shiva-Reddy SG, et al. Carbon nanotube based high resolution holograms. Advanced Materials. 2012;**24**:OP331-OP336. DOI: 10.1002/adma.201202593

[45] Walther B, Helgert C, Rockstuhl C, Setzpfandt F, Eilenberger F, Kley E, et al. Photonics: Spatial and spectral light shaping with metamaterials (Adv. Mater. 47/2012). Advanced Materials. 2012;**24**:6251-6251. DOI: 10.1002/ adma.201290300

[46] Overvig AC, Shrestha S, Malek SC, Lu M, Stein A, Zheng C, et al. Dielectric metasurfaces for complete and independent control of the optical amplitude and phase. Light: Science & Applications. 2019;**8**:92. DOI: 10.1038/s41377-019-0201-7

[47] Ni X, Kildishev AV, Shalaev VM. Metasurface holograms for visible light. Nature Communications. 2013;**4**:2807. DOI: 10.1038/ncomms3807

[48] Wang Q, Zhang X, Xu Y, Gu J, Li Y, Tian Z, et al. Broadband metasurface holograms: Toward complete phase and amplitude engineering. Scientific Reports. 2016;**6**:32867. DOI: 10.1038/srep32867

[49] Chong KE, Wang L, Staude I, James AR, Dominguez J, Liu S, et al. Efficient polarization-insensitive complex wavefront control using Huygens' metasurfaces based on dielectric resonant meta-atoms. ACS Photonics. 2016;**3**:514-519. DOI: 10.1021/acsphotonics.5b00678

[50] Gao H, Fan X, Xiong W, Hong M. Recent advances in optical dynamic meta-holography. Opto-Electronic Advances. 2021;**4**:210030. DOI: 10.29026/oea.2021.210030

[51] Jiang Q, Jin G, Cao L. When metasurface meets hologram: Principle and advances. Advances in Optics and Photonics. 2019;**11**:518. DOI: 10.1364/AOP.11.000518

[52] Chen S, Liu W, Li Z, Cheng H, Tian J. Metasurface-empowered optical multiplexing and multifunction. Advanced Materials. 2020;**32**:1805912. DOI: 10.1002/adma.201805912

[53] Li Z, Yu S, Zheng G. Advances in exploiting the degrees of freedom in nanostructured metasurface design: From 1 to 3 to more. Nano. 2020;**9**: 3699-3731. DOI: 10.1515/nanoph-2020-0127

[54] Cui T, Bai B, Sun H. Tunable metasurfaces based on active materials. Advanced Functional Materials. 2019;**29**: 1806692. DOI: 10.1002/adfm.201806692

[55] Nemati A, Wang Q, Hong M, Teng J. Tunable and reconfigurable metasurfaces and metadevices. Opto-Electronic Advances. 2018;**1**:180009. DOI: 10.29026/oea.2018.180009

[56] Horie Y, Arbabi A, Arbabi E, Kamali SM, Faraon A. High-speed, phase-dominant spatial light modulation with silicon-based active resonant antennas. ACS Photonics. 2018;**5**: 1711-1717. DOI: 10.1021/acsphotonics.7b01073

[57] Sun J, Timurdogan E, Yaacobi A, Hosseini ES, Watts MR. Large-scale nanophotonic phased array. Nature. 2013;**493**:195-199. DOI: 10.1038/nature11727

[58] Rahmani M, Xu L, Miroshnichenko AE, Komar A, Camacho-Morales R, Chen H, et al. Reversible thermal tuning of all-dielectric metasurfaces. Advanced Functional Materials. 2017;**27**:1700580. DOI: 10.1002/adfm.201700580

[59] Lewi T, Evans HA, Butakov NA, Schuller JA. Ultrawide thermo-optic tuning of PbTe meta-atoms. Nano Letters. 2017;**17**:3940-3945. DOI: 10.1021/acs.nanolett.7b01529

[60] Gu J, Singh R, Liu X, Zhang X, Ma Y, Zhang S, et al. Active control of electromagnetically induced transparency analogue in terahertz metamaterials. Nature Communications. 2012;**3**:1151. DOI: 10.1038/ncomms2153

[61] Makarov S, Kudryashov S, Mukhin I, Mozharov A, Milichko V, Krasnok A, et al. Tuning of magnetic optical response in a dielectric nanoparticle by ultrafast photoexcitation of dense electron–hole plasma. Nano Letters. 2015;**15**:6187-6192. DOI: 10.1021/acs.nanolett.5b02534

[62] Lewi T, Iyer PP, Butakov NA, Mikhailovsky AA, Schuller JA. Widely tunable infrared antennas using free carrier refraction. Nano Letters. 2015;**15**: 8188-8193. DOI: 10.1021/acs.nanolett.5b03679

[63] Chen HT, Padilla WJ, Zide JMO, Gossard AC, Taylor AJ, Averitt RD. Active terahertz metamaterial devices. Nature. 2006;**444**:597-600. DOI: 10.1038/nature05343

[64] Watts CM, Shrekenhamer D, Montoya J, Lipworth G, Hunt J, Sleasman T, et al. Terahertz compressive imaging with metamaterial spatial light modulators. Nature Photonics. 2014;**8**: 605-609. DOI: 10.1038/nphoton.2014.139

[65] Chen HT, Padilla WJ, Cich MJ, Azad AK, Averitt RD, Taylor AJ. A metamaterial solid-state terahertz phase modulator. Nature Photonics. 2009;**3**:148-151. DOI: 10.1038/nphoton.2009.3

[66] Iyer PP, Pendharkar M, Schuller JA. Electrically reconfigurable metasurfaces using heterojunction resonators. Advanced Optical Materials. 2016;**4**: 1582-1588. DOI: 10.1002/adom.201600297

[67] Huang YW, Lee HWH, Sokhoyan R, Pala RA, Thyagarajan K, Han S, et al. Gate-tunable conducting oxide metasurfaces. Nano Letters. 2016;**16**: 5319-5325. DOI: 10.1021/acs.nanolett.6b00555

[68] Chen Y, Ke F, Ci P, Ko C, Park T, Saremi S, et al. Pressurizing field-effect transistors of few-layer MoS_2 in a diamond anvil cell. Nano Letters. 2017; **17**:194-199. DOI: 10.1021/acs.nanolett.6b03785

[69] Thyagarajan K, Sokhoyan R, Zornberg L, Atwater HA. Millivolt modulation of plasmonic metasurface optical response via ionic conductance. Advanced Materials. 2017;**29**:1701044. DOI: 10.1002/adma.201701044

[70] Shirmanesh GK, Sokhoyan R, Wu PC, Atwater HA. Electro-optically tunable multifunctional metasurfaces. ACS Nano. 2020;**14**:6912-6920. DOI: 10.1021/acsnano.0c01269

[71] Lee S, Baek S, Kim T, Cho H, Lee S, Kang J, et al. Metamaterials for enhanced optical responses and their application to active control of terahertz waves. Advanced Materials. 2020;**32**:2000250. DOI: 10.1002/adma.202000250

[72] Liu PQ, Luxmoore IJ, Mikhailov SA, Savostianova NA, Valmorra F, Faist J, et al. Highly tunable hybrid metamaterials employing split-ring resonators strongly coupled to graphene surface plasmons. Nature Communications. 2015;**6**:8969. DOI: 10.1038/ncomms9969

[73] Li X, Ren H, Chen X, Liu J, Li Q, Li C, et al. Athermally photoreduced graphene oxides for three-dimensional holographic images. Nature Communications. 2015;**6**:6984. DOI: 10.1038/ncomms7984

[74] Li SQ, Xu X, Maruthiyodan Veetil R, Valuckas V, Paniagua-Domínguez R, Kuznetsov AI. Phase-only transmissive spatial light modulator based on tunable dielectric metasurface. Science. 2019; **364**:1087-1090. DOI: 10.1126/science.aaw6747

[75] Komar A, Paniagua-Domínguez R, Miroshnichenko A, Yu YF, Kivshar YS, Kuznetsov AI, et al. Dynamic beam switching by liquid crystal tunable dielectric metasurfaces. ACS Photonics. 2018;**5**:1742-1748. DOI: 10.1021/acsphotonics.7b01343

[76] De Galarreta CR, Alexeev AM, Au Y, Lopez-Garcia M, Klemm M, Cryan M, et al. Nonvolatile reconfigurable phase-change metadevices for beam steering in the near infrared. Advanced Functional Materials. 2018;**28**:1704993. DOI: 10.1002/adfm.201704993

[77] Yin X, Steinle T, Huang L, Taubner T, Wuttig M, Zentgraf T, et al. Beam switching and bifocal zoom lensing using active plasmonic metasurfaces. Light: Science & Applications. 2017;**6**:e17016-e17016. DOI: 10.1038/lsa.2017.16

[78] Zhou H, Wang Y, Li X, Wang Q, Wei Q, Geng G, et al. Switchable active phase modulation and holography encryption based on hybrid metasurfaces. Nano. 2020;**9**:905-912. DOI: 10.1515/nanoph-2019-0519

[79] Lee SY, Kim YH, Cho SM, Kim GH, Kim TY, Ryu H, et al. Holographic image generation with a thin-film resonance caused by chalcogenide phase-change material. Scientific Reports. 2017;**7**:41152. DOI: 10.1038/srep41152

[80] Zhang M, Pu M, Zhang F, Guo Y, He Q, Ma X, et al. Plasmonic metasurfaces for switchable photonic spin–orbit interactions based on phase change materials. Advanced Science. 2018;**5**:1800835. DOI: 10.1002/advs.201800835

[81] Raeis-Hosseini N, Rho J. Metasurfaces based on phase-change material as a reconfigurable platform for multifunctional devices. Materials. 2017;**10**:1046. DOI: 10.3390/ma10091046

[82] Qu Y, Li Q, Du K, Cai L, Lu J, Qiu M. Dynamic thermal emission control based on ultrathin plasmonic metamaterials including phase-changing material GST. Laser & Photonics Reviews. 2017;**11**:1700091. DOI: 10.1002/lpor.201700091

[83] Driscoll T, Palit S, Qazilbash MM, Brehm M, Keilmann F, Chae BG, et al. Dynamic tuning of an infrared hybrid-metamaterial resonance using vanadium dioxide. Applied Physics Letters. 2008;**93**:024101. DOI: 10.1063/1.2956675

[84] Goldflam MD, Liu MK, Chapler BC, Stinson HT, Sternbach AJ, McLeod AS, et al. Voltage switching of a VO_2 memory metasurface using ionic gel. Applied Physics Letters. 2014;**105**:041117. DOI: 10.1063/1.4891765

[85] Liu X, Wang Q, Zhang X, Li H, Xu Q, Xu Y, et al. Thermally dependent dynamic meta-holography using a vanadium dioxide integrated metasurface. Advanced Optical Materials. 2019;**7**:1900175. DOI: 10.1002/adom.201900175

[86] Haimov T, Aydin K, Scheuer J. Reconfigurable holograms using VO_2 -based tunable metasurface. IEEE Journal of Selected Topics in Quantum Electronics. 2021;**27**:1-8. DOI: 10.1109/JSTQE.2020.3011678

[87] Song S, Ma X, Pu M, Li X, Liu K, Gao P, et al. Actively tunable structural color rendering with tensile substrate. Advanced Optical Materials. 2017;**5**:1600829. DOI: 10.1002/adom.201600829

[88] Zhang C, Jing J, Wu Y, Fan Y, Yang W, Wang S, et al. Stretchable all-dielectric metasurfaces with polarization-insensitive and full-

spectrum response. ACS Nano. 2020;**14**: 1418-1426. DOI: 10.1021/acsnano.9b08228

[89] Malek SC, Ee HS, Agarwal R. Strain multiplexed metasurface holograms on a stretchable substrate. Nano Letters. 2017;**17**:3641-3645. DOI: 10.1021/acs.nanolett.7b00807

[90] Ee HS, Agarwal R. Tunable metasurface and flat optical zoom lens on a stretchable substrate. Nano Letters. 2016;**16**:2818-2823. DOI: 10.1021/acs.nanolett.6b00618

[91] Li T, Wei Q, Reineke B, Walter F, Wang Y, Zentgraf T, et al. Reconfigurable metasurface hologram by utilizing addressable dynamic pixels. Optics Express. 2019;**27**:21153. DOI: 10.1364/OE.27.021153

[92] Li J, Kamin S, Zheng G, Neubrech F, Zhang S, Liu N. Addressable metasurfaces for dynamic holography and optical information encryption. Science Advances. 2018;**4**:eaar6768. DOI: 10.1126/sciadv.aar6768

[93] Duan X, Kamin S, Sterl F, Giessen H, Liu N. Hydrogen-regulated chiral nanoplasmonics. Nano Letters. 2016;**16**: 1462-1466. DOI: 10.1021/acs.nanolett.5b05105

[94] Lin SF, Kim ES. Single SLM full-color holographic 3-D display based on sampling and selective frequency-filtering methods. Optics Express. 2017; **25**:11389. DOI: 10.1364/OE.25.011389

[95] Jesacher A, Bernet S, Ritsch-Marte M. Colour hologram projection with an SLM by exploiting its full phase modulation range. Optics Express. 2014; **22**:20530. DOI: 10.1364/OE.22.020530

[96] Huang YW, Chen WT, Tsai WY, Wu PC, Wang CM, Sun G, et al. Aluminum plasmonic multicolor meta-hologram. Nano Letters. 2015;**15**: 3122-3127. DOI: 10.1021/acs.nanolett.5b00184

[97] Wang B, Dong F, Li QT, Yang D, Sun C, Chen J, et al. Visible-frequency dielectric metasurfaces for multiwavelength achromatic and highly dispersive holograms. Nano Letters. 2016;**16**:5235-5240. DOI: 10.1021/acs.nanolett.6b02326

[98] Zhao W, Liu B, Jiang H, Song J, Pei Y, Jiang Y. Full-color hologram using spatial multiplexing of dielectric metasurface. Optics Letters. 2016;**41**:147. DOI: 10.1364/OL.41.000147

[99] Li X, Chen L, Li Y, Zhang X, Pu M, Zhao Z, et al. Multicolor 3D meta-holography by broadband plasmonic modulation. Science Advances. 2016;**2**: e1601102. DOI: 10.1126/sciadv.1601102

[100] Wan W, Gao J, Yang X. Full-color plasmonic metasurface holograms. ACS Nano. 2016;**10**:10671-10680. DOI: 10.1021/acsnano.6b05453

[101] Zhang X, Pu M, Guo Y, Jin J, Li X, Ma X, et al. Colorful metahologram with independently controlled images in transmission and reflection spaces. Advanced Functional Materials. 2019;**29**: 1809145. DOI: 10.1002/adfm.201809145

[102] Montelongo Y, Tenorio-Pearl JO, Milne WI, Wilkinson TD. Polarization switchable diffraction based on subwavelength plasmonic nanoantennas. Nano Letters. 2014;**14**:294-298. DOI: 10.1021/nl4039967

[103] Chen WT, Yang KY, Wang CM, Huang YW, Sun G, Chiang ID, et al. High-efficiency broadband meta-hologram with polarization-controlled dual images. Nano Letters. 2014;**14**: 225-230. DOI: 10.1021/nl403811d

[104] Zhang F, Pu M, Li X, Gao P, Ma X, Luo J, et al. All-dielectric metasurfaces for simultaneous giant circular asymmetric transmission and wavefront shaping based on asymmetric photonic spin–orbit interactions. Advanced Functional Materials. 2017;**27**:1704295. DOI: 10.1002/adfm.201704295

[105] Deng ZL, Deng J, Zhuang X, Wang S, Li K, Wang Y, et al. Diatomic metasurface for vectorial holography. Nano Letters. 2018;**18**:2885-2892. DOI: 10.1021/acs.nanolett.8b00047

[106] Wang Q, Plum E, Yang Q, Zhang X, Xu Q, Xu Y, et al. Reflective chiral meta-holography: Multiplexing holograms for circularly polarized waves. Light: Science & Applications. 2018;**7**:25. DOI: 10.1038/s41377-018-0019-8

[107] Wen D, Yue F, Li G, Zheng G, Chan K, Chen S, et al. Helicity multiplexed broadband metasurface holograms. Nature Communications. 2015;**6**:8241. DOI: 10.1038/ncomms9241

[108] Arbabi A, Horie Y, Bagheri M, Faraon A. Dielectric metasurfaces for complete control of phase and polarization with subwavelength spatial resolution and high transmission. Nature Nanotechnology. 2015;**10**:937-943. DOI: 10.1038/nnano.2015.186

[109] Balthasar Mueller JP, Rubin NA, Devlin RC, Groever B, Capasso F. Metasurface polarization optics: Independent phase control of arbitrary orthogonal states of polarization. Physical Review Letters. 2017;**118**: 113901. DOI: 10.1103/PhysRevLett.118.113901

[110] Khorasaninejad M, Ambrosio A, Kanhaiya P, Capasso F. Broadband and chiral binary dielectric meta-holograms. Science Advances. 2016;**2**:e1501258. DOI: 10.1126/sciadv.1501258

[111] Arbabi E, Kamali SM, Arbabi A, Faraon A. Vectorial holograms with a dielectric metasurface: Ultimate polarization pattern generation. ACS Photonics. 2019;**6**:2712-2718. DOI: 10.1021/acsphotonics.9b00678

[112] Deng L, Deng J, Guan Z, Tao J, Chen Y, Yang Y, et al. Malus-metasurface-assisted polarization multiplexing. Light: Science & Applications. 2020;**9**:101. DOI: 10.1038/s41377-020-0327-7

[113] Zhao R, Sain B, Wei Q, Tang C, Li X, Weiss T, et al. Multichannel vectorial holographic display and encryption. Light: Science & Applications. 2018;**7**:95. DOI: 10.1038/s41377-018-0091-0

[114] Kamali SM, Arbabi E, Arbabi A, Horie Y, Faraji-Dana M, Faraon A. Angle-multiplexed metasurfaces: Encoding independent wavefronts in a single metasurface under different illumination angles. Physical Review X. 2017;**7**:041056. DOI: 10.1103/PhysRevX.7.041056

[115] Wang E, Niu J, Liang Y, Li H, Hua Y, Shi L, et al. Complete control of multichannel, angle-multiplexed, and arbitrary spatially varying polarization fields. Advanced Optical Materials. 2020;**8**:1901674. DOI: 10.1002/adom.201901674

[116] Zhang X, Jin J, Pu M, Li X, Ma X, Gao P, et al. Ultrahigh-capacity dynamic holographic displays via anisotropic nanoholes. Nanoscale. 2017;**9**:1409-1415. DOI: 10.1039/C6NR07854K

[117] Ren H, Briere G, Fang X, Ni P, Sawant R, Héron S, et al. Metasurface orbital angular momentum holography.

Nature Communications. 2019;**10**:2986. DOI: 10.1038/s41467-019-11030-1

[118] Jin L, Huang YW, Jin Z, Devlin RC, Dong Z, Mei S, et al. Dielectric multi-momentum meta-transformer in the visible. Nature Communications. 2019; **10**:4789. DOI: 10.1038/s41467-019-12637-0

[119] Zhou H, Sain B, Wang Y, Schlickriede C, Zhao R, Zhang X, et al. Polarization-encrypted orbital angular momentum multiplexed metasurface holography. ACS Nano. 2020;**14**: 5553-5559. DOI: 10.1021/acsnano.9b09814

[120] Yu P, Li J, Li X, Schütz G, Hirscher M, Zhang S, et al. Generation of switchable singular beams with dynamic metasurfaces. ACS Nano. 2019;**13**: 7100-7106. DOI: 10.1021/acsnano.9b02425

[121] Fang X, Ren H, Gu M. Orbital angular momentum holography for high-security encryption. Nature Photonics. 2020;**14**:102-108. DOI: 10.1038/s41566-019-0560-x

[122] Ren H, Fang X, Jang J, Bürger J, Rho J, Maier SA. Complex-amplitude metasurface-based orbital angular momentum holography in momentum space. Nature Nanotechnology. 2020;**15**: 948-955. DOI: 10.1038/s41565-020-0768-4

[123] Izumi R, Ikezawa S, Iwami K. Metasurface holographic movie: A cinematographic approach. Optics Express. 2020;**28**:23761. DOI: 10.1364/OE.399369

[124] Gao H, Wang Y, Fan X, Jiao B, Li T, Shang C, et al. Dynamic 3D meta-holography in visible range with large frame number and high frame rate. Science Advances. 2020;**6**:eaba8595. DOI: 10.1126/sciadv.aba8595

[125] Liu Y, Xu K, Fan X, Wang X, Yu X, Xiong W, et al. Dynamic interactive bitwise meta-holography with ultra-high computational and display frame rates. Opto-Electronic Advances. 2024;**7**: 230108. DOI: 10.29026/oea.2024.230108

[126] Xu K, Liu Y, Fan X, Wang X, Yu X, Teng Z, et al. Simultaneous dynamic display of meta-hologram and meta-nanoprinting with high frame rate. Laser & Photonics Reviews. 2025;**19**:2400815. DOI: 10.1002/lpor.202400815

[127] Li G, Chen S, Pholchai N, Reineke B, Wong PWH, Pun EYB, et al. Continuous control of the nonlinearity phase for harmonic generations. Nature Materials. 2015;**14**:607-612. DOI: 10.1038/nmat4267

[128] Segal N, Keren-Zur S, Hendler N, Ellenbogen T. Controlling light with metamaterial-based nonlinear photonic crystals. Nature Photonics. 2015;**9**: 180-184. DOI: 10.1038/nphoton.2015.17

[129] Tymchenko M, Gomez-Diaz JS, Lee J, Nookala N, Belkin MA, Alù A. Gradient nonlinear Pancharatnam-Berry metasurfaces. Physical Review Letters. 2015;**115**:207403. DOI: 10.1103/PhysRevLett.115.207403

[130] Ye W, Zeuner F, Li X, Reineke B, He S, Qiu CW, et al. Spin and wavelength multiplexed nonlinear metasurface holography. Nature Communications. 2016;**7**:11930. DOI: 10.1038/ncomms11930

[131] Qu G, Yang W, Song Q, Liu Y, Qiu CW, Han J, et al. Reprogrammable meta-hologram for optical encryption. Nature Communications. 2020;**11**:5484. DOI: 10.1038/s41467-020-19312-9

[132] Jin L, Dong Z, Mei S, Yu YF, Wei Z, Pan Z, et al. Noninterleaved metasurface for (2^6-1) spin- and wavelength-encoded holograms. Nano Letters. 2018;**18**: 8016-8024. DOI: 10.1021/acs.nanolett.8b04246

[133] Liang C, Wang J, Huang T, Dai Q, Li Z, Yu S, et al. Structural-color meta-nanoprinting embedding multi-domain spatial light field information. Nano. 2024;**13**:1665-1675. DOI: 10.1515/nanoph-2024-0019

[134] Deng Z, Jin M, Ye X, Wang S, Shi T, Deng J, et al. Full-color complex-amplitude vectorial holograms based on multi-freedom metasurfaces. Advanced Functional Materials. 2020;**30**:1910610. DOI: 10.1002/adfm.201910610

[135] Li Z, Chen C, Guan Z, Tao J, Chang S, Dai Q, et al. Three-channel metasurfaces for simultaneous meta-holography and meta-nanoprinting: A single-cell design approach. Laser & Photonics Reviews. 2020;**14**:2000032. DOI: 10.1002/lpor.202000032

[136] Dai Q, Guan Z, Chang S, Deng L, Tao J, Li Z, et al. A single-celled tri-functional metasurface enabled with triple manipulations of light. Advanced Functional Materials. 2020;**30**:2003990. DOI: 10.1002/adfm.202003990

[137] Gao H, Fan X, Wang Y, Liu Y, Wang X, Xu K, et al. Multi-foci metalens for spectra and polarization ellipticity recognition and reconstruction. Opto-Electronic Science. 2023;**2**:220026. DOI: 10.29026/oes.2023.220026

[138] Gao H, Fan X, Wang Y, Wang X, Xu K, Zeng C, et al. Metasurface-based orbital angular momentum multi-dimensional demultiplexer and decoder. Laser & Photonics Reviews. 2023;**17**:2300393. DOI: 10.1002/lpor.202300393

[139] Wang X, Fan X, Liu Y, Xu K, Zhou Y, Zhang Z, et al. 3D nanolithography via holographic multi-focus metalens. Laser & Photonics Reviews. 2024;**18**:2400181. DOI: 10.1002/lpor.202400181

Chapter 3

Optical Tools for Holographic Lens Design in Anidolic Solar Applications

Tomás Lloret, Marta Morales-Vidal and Inmaculada Pascual

Abstract

This book chapter provides an overview of how holographic lenses (HLs) applied as anidolic systems (solar concentrators) have evolved to optimize the characteristics of current photovoltaic devices. The main parameters to be considered in the design of these HLs are described, such as average diffraction efficiency, acceptance angle, angular and spectral dependence, and the optimal number of HLs. In addition, the dynamic range of the material and the characteristics to multiplex different HLs with high diffraction efficiencies are emphasized. To this end, new strategies are proposed, such as the use of the exposure scheduling method (ESM) to improve the exposure times. This technique is compared to current techniques followed by researchers working in this field, such as the stacking of several photopolymer layers or considering a system of N holographic lenses. Finally, the advantages and disadvantages of current holographic photopolymers are compared.

Keywords: holographic lenses, photopolymers, multiplexing, dynamic range, solar energy, photovoltaics

1. Introduction

Sunlight can be harnessed directly in two ways: by heating a fluid (solar thermal energy) or by generating an electric current (solar photovoltaics). In this chapter, we will focus on the second mode. The main advantage of this technology is that it is clean energy, i.e. it does not generate polluting waste like nuclear energy or fuel-based energy and focuses on the manufacturing process of photovoltaic panels and solar concentrators. Solar photovoltaic conversion is an increasingly important source of clean energy that can reduce the effects of climate change. However, these systems currently provide only a small fraction of total global energy production. For them to become a more significant energy source, the efficiency of PV conversion needs to be increased with lower-cost systems. Optics can contribute to meeting these requirements by performing light management operations that improve the efficiency of PV systems. Nowadays, concentrator photovoltaic technology is increasingly recognized as the most promising technology to meet the energy challenges facing the world and holography is a valuable aim to make it possible. Holographic optical elements are particularly suited to this task, as they can be used to realize highly functional components in thin film materials and manufactured using high-volume production methods to reduce cost.

Solar concentrators are optical elements that can normally be based on the phenomena of refraction and reflection and whose objective is to redirect sunlight toward the photovoltaic cell in order to increase efficiency. The most commonly used concentrators are Fresnel lenses, parabolic mirrors, and reflectors [1]. The optical-geometric scheme of these elements can be seen in **Figure 1**. Some advantages and disadvantages of these elements are as follows [2]:

- Its use allows to increase the energy flux reaching the photovoltaic cell.

- Fresnel lenses have a small acceptance angle and therefore require the use of solar trackers.

- Parabolic mirrors and reflectors have a wider acceptance angle than Fresnel lenses, but these elements tend to deteriorate over time as they deteriorate due to erosion.

Researchers are trying to solve the problem of using solar trackers by designing holographic solar concentrators [3, 4]. Holographic solar concentrators (HSCs) are concentrators based on holographic optical elements (HOEs) that can act by

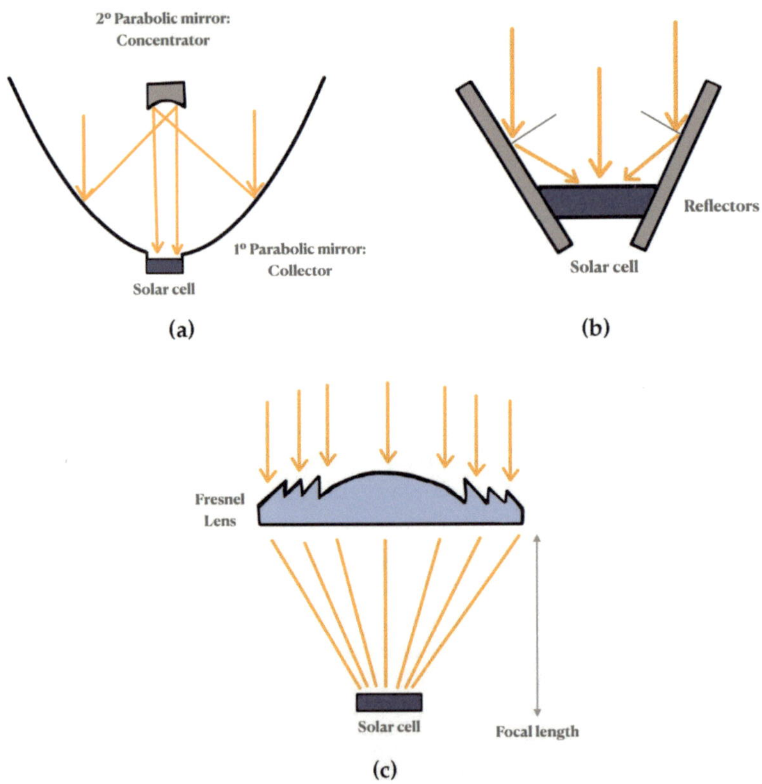

Figure 1.
Most common conventional solar concentrators: (a) parabolic mirrors, (b) reflectors, and (c) Fresnel lenses.

transmission or reflection [5–10]. Holographic optical elements (HOEs) are optical elements obtained from the holographic process. Although its principle of operation is based on the phenomenon of diffraction, it can simulate the behavior of conventional optical elements, such as holographic lenses (HLs). Some advantages of HOEs over conventional optical elements are as follows:

- The physical structure of HOEs is the recording material layer, which allows the optical element to be attached to different types of media.
- Different types of HOEs can be manufactured on the same recording material layer, in case different functions are required at the same time, etc.
- Reduced weight and volume, as well as ease of mass production by copying methods. This results in reduced manufacturing costs, which is a significant advantage.

Usually, the most famous holographic concentrators are HLs [11]. These lenses are holograms of an object point located at a certain distance from the recording material (**Figure 2**). HLs are obtained from the interference of a plane beam (reference beam) and a convergent or divergent beam (object beam). Usually, convergent refractive lenses or microscope objectives are used to obtain the object beam.

Unlike refractive lenses, HLs have spectral and angular dependence [12]. It should be noted that the spectral dependence of HLs is inverse to the spectral dependence of refractive lenses. These dependencies are described by the Eqs. (1) and (2), respectively. In addition, the dependence of the focal length on the reconstruction wavelength can be seen in **Figure 3**.

$$\frac{1}{f'_{HL}} = \frac{1}{R_c} \pm \mu \left(\frac{1}{R_o} - \frac{1}{R_r} \right) \tag{1}$$

$$|\sin(\theta_i)| = |\sin(\theta_c)| \pm \mu \left(|\sin(\theta_o)| - |\sin(\theta_r)| \right) \tag{2}$$

where i, c, o, r are the subscripts of the image, reconstruction, object, and reference beams, respectively; and R is the distance from the source point to the hologram. In addition, considering that the reference and reconstruction beams are plane waves, R_r and R_c tend to infinity, and R_o is the refraction focal length. Furthermore, f'_{HL} is the

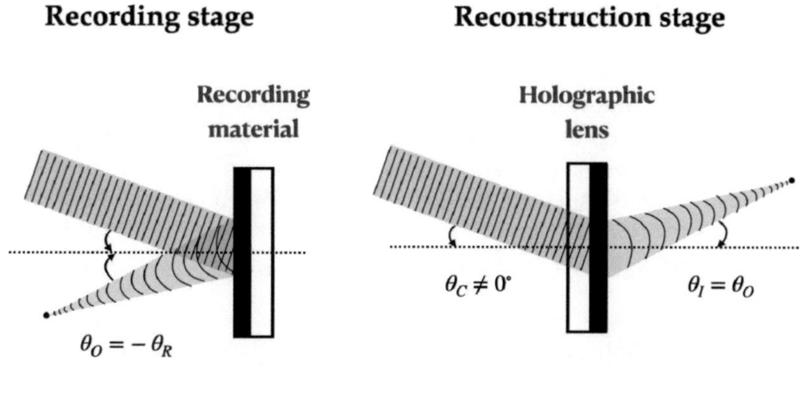

Figure 2.
Example of the holographic process for a transmission negative, off-axis, holographic lens with symmetrical recording.

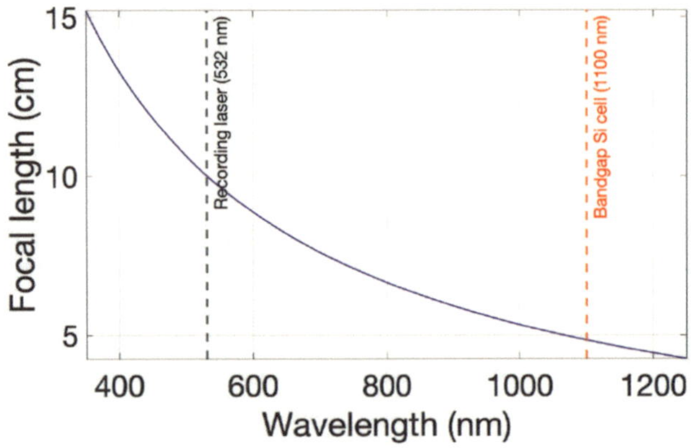

Figure 3.
Example of the relationship between the focal length and the reconstruction wavelength, for an HL recorded with an $f'_{HL} = 10$ cm *with a laser emitting at 532 nm.*

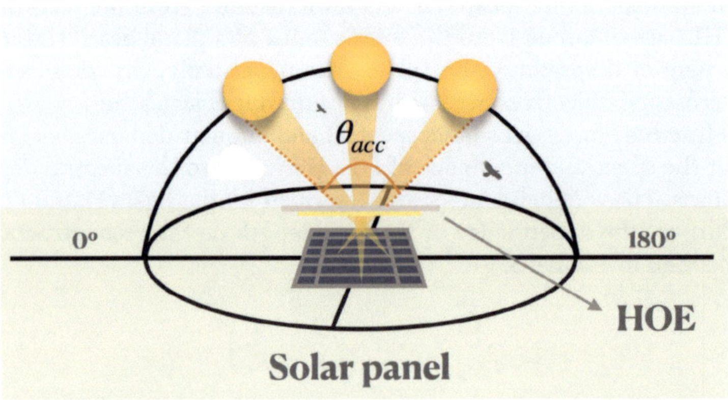

Figure 4.
Illustration showing the definition of acceptance angle for an HSC operating on a solar panel.

focal length of each lens, while μ is the ratio of the reconstruction wavelength to the recording wavelength $\mu = \lambda_c/\lambda_0$.

2. Acceptance angle (θ_{acc})

The acceptance angle (θ_{acc}) is a parameter that is defined to study the behavior of concentrators in real conditions or when working with solar simulators. It is the maximum angle for which the concentrator is able to concentrate sunlight on the cell. The concept of acceptance angle can be seen in **Figure 4**. These standard concentrators operate at restricted angles of incidence, so collecting a large acceptance angle at the cell requires continuous orientation of the sun with expensive solar trackers, which can also affect the land of crucial ecosystems [2].

3. Diffraction efficiency (η)

The tool or parameter used par excellence to characterize diffraction gratings and HLs is the diffraction efficiency. Diffraction efficiency (η) describes the amount of optical power diffracted by a hologram at a desired wavefront [13]. Specifically, it is defined as the diffracted irradiance (E_d) in one or more diffraction orders relative to the irradiance (E_i) incident on the hologram. According to this definition, we have that

$$\eta_i = \frac{E_d}{E_i} = \frac{P_d/A}{P_i/A} = \frac{P_d}{P_i} \tag{3}$$

where P_d is the diffracted optical power at the order, P_i is the power incident on the hologram, and A is the surface area of the photodetector. To avoid having to make corrections in the calculation of η (absorption losses, Fresnel reflections, etc.), a good method is to consider the following definition

$$\eta_i = \frac{P_d}{P_d + P_t} \tag{4}$$

where P_t is the transmitted power, so instead of considering the incident power, the sum of the diffracted and transmitted power is considered. In this chapter, we have used this definition to analyze the diffraction efficiency experimentally.

Most current holographic solar concentrator designs have more than one holographic lens, so the average diffraction efficiency must be defined. For this purpose, depending on the holographic solar concentrator design, one definition or another is used.

3.1 Theoretical study based on Kogelnik's coupled wave analysis (KCWA)

Diffraction efficiency can be studied theoretically in different ways. In this work, we have used Kogelnik's coupled wave analysis (KCWA), which is an approximate theory with a simple analytical solution [13, 14]. This theory has been used to follow different strategies in the design of holographic concentrators and to study how the diffraction efficiency and the acceptance angle can be increased. This approach provides a prediction of the diffraction efficiency at a specific wavelength, λ_C, near the Bragg angle for a transmission phase grating with a given thickness, d, we have that

$$\eta = \frac{\sin^2\left(\xi^2 + v^2\right)^{\frac{1}{2}}}{1 + \xi^2/v^2} \tag{5}$$

where in this equation, the variable ξ varies according to the detuning parameter (ϑ), as described in Eqs. (6) and (7). This dependency is influenced by the grating period (Λ), the reconstruction angle in air (θ_c), the reconstruction angle within the material (θ'_c), the reconstruction wavelength (λ_c), the fringe slope (φ), and the refractive index (n). The value of ξ is zero under Bragg conditions and becomes either positive or negative when the system deviates from Bragg conditions. For the lenses, the fit was performed considering only the central value of the grating vector (K) due to its small diameter,

$$\vartheta = K\left[\cos(\varphi - \theta_c) - \frac{K}{4\pi n}\lambda_c\right] \tag{6}$$

$$\xi = \frac{\vartheta d}{2c_s} \tag{7}$$

In addition, the grating strength, v, is defined as a function of the coefficients c_r and c_s. It is important to note that the coefficient c_s is different for symmetric recordings (where $\varphi = \pi/2$) and asymmetric recordings (where $\varphi \neq \pi/2$).

$$v = \frac{\kappa d}{\sqrt{c_r c_s}} \tag{8}$$

where $c_r = \cos(\theta)$ and $c_s = \cos(\theta) - \frac{K}{\beta}\cos\varphi$.
Detailed study by means of KCWA has revealed the following aspects:

- It can be observed that when the spatial frequency decreases, the angular width increases, so it is interesting to design low spatial frequency holographic concentrators.

- It can also be observed that when the optical thickness decreases, the angular width increases, but the diffraction efficiency decreases, so depending on the technique used, it will be better to use more or less thickness. For example, in the case of the multiplexing technique, it is important that the thickness is thicker in order to have the capacity to store more holograms.

Figure 5 shows the Bragg angular selectivity curves that justify the previous points.

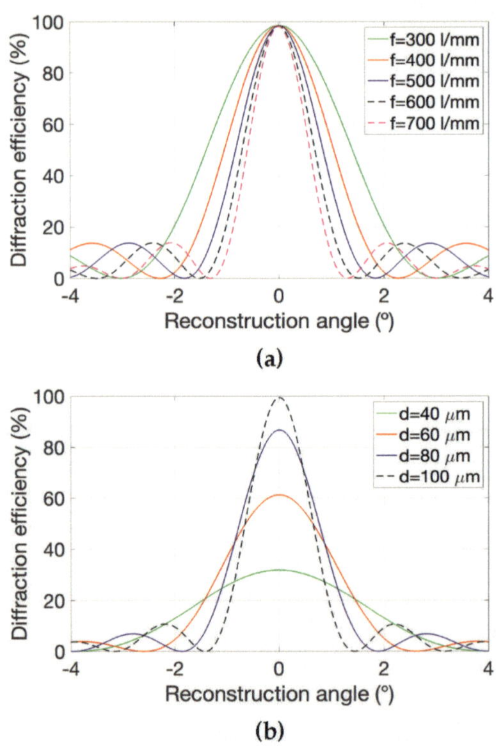

Figure 5.
Theoretical study based on KCWA. (a) Influence of spatial frequency on diffraction efficiency and angular width. (b) Influence of optical thickness on diffraction efficiency and angular width.

3.2 Average diffraction efficiency (η_{AVR})

Most designs for holographic solar concentrators are based on more than one hologram. In case the system is composed of several holographic optical elements, the average diffraction efficiency (η_{avr}) must be defined to get an overall idea of the efficiency of the final design [13]. This average diffraction efficiency is defined differently depending on the technique used. In this work, the multiplexing technique has been used to obtain holographic solar concentrators based on angularly multiplexed holographic lenses.

3.2.1 Holographic solar concentrators based on holographic N-lenses

The simplest holographic solar concentrator design is a system based on $1, 2, \ldots, N$ holographic lenses, each lens working for a given illumination angle. In this case, the average diffraction efficiency is calculated by averaging the diffraction efficiencies. The mathematical definition is as follows

$$\eta_{avr} = \frac{\eta_1 + \eta_2 + \cdots + \eta_N}{N} = \frac{1}{N}\sum_{i=1}^{N} \eta_i \tag{9}$$

where N is the number of holographic lenses that make up the concentrator system.

3.2.2 Multiplexing-based holographic solar concentrators

To determine the average diffraction efficiency of multiplex-based HSCs, the dynamic range must first be calculated [7, 15]. The dynamic range is defined as the number of holograms with diffraction efficiency $\eta = 1$ (100%) that can be stored in a material with a given thickness. Its mathematical expression is given by

$$M\# = \sum_{i=1}^{N} (\eta_i)^{1/2} \tag{10}$$

where η_i is the diffraction efficiency of each hologram, and N is the number of recorded holograms. To calculate the dynamic range, we must sum the square roots of the diffraction efficiencies of all holograms that have been recorded in the material. From the dynamic range, it is possible to determine the number of holograms that can be stored in the material to obtain a certain diffraction efficiency, or alternatively the average diffraction efficiency for a given number of stored holograms. The latter average diffraction efficiency is given by

$$\eta_{avr} = \left(\frac{M\#}{N}\right)^2 \tag{11}$$

3.2.3 Stacking-based holographic solar concentrators

On the other hand, for holographic concentrators based on the stacking technique, the average diffraction efficiency must be calculated as the product of the diffraction efficiencies of each of the layers that make up the design. The mathematical definition would be as follows

$$\eta_{avr} = \eta_1 \cdot \eta_2 \cdot \ \dots \ \cdot \eta_N = \prod_{i=1}^{N} \eta_i \tag{12}$$

3.3 Exposure schedule method (ESM)

The first step in using ESM is to characterize the recording material with a specific hologram exposure method [13, 16, 17]. In this work, the angular multiplexing technique has been used with a certain delay time between each exposure on a specific type of photopolymer. This is necessary because ESM and delay time affect the monomer diffusion mechanism of the hologram formation process. The next step in the development of an exposure program is to approximate the characteristic curve for the refractive index modulation (n_1) with a polynomial function defined as

$$A = a_0 + a_1E + a_2E^2 + a_3E^3 + a_4E^4 + a_5E^5 + a_6E^6 \tag{13}$$

where A is the cumulative grating intensity for N holograms, E is the cumulative exposure energy, and a_i are the polynomial coefficients determined from the curve fitting function. It can be seen that the derivative with respect to the exposure energy defines the rate of change of the grating resistance with exposure energy. Therefore, if the entire dynamic range of the material is divided equally among all multiplexed holograms, the required exposure schedule is defined as

$$\frac{A_{Sat}}{N} = \left[\frac{\partial A}{\partial E}\right]_{E=\sum_{i=1}^{n-1} E_i} \cdot E_n \tag{14}$$

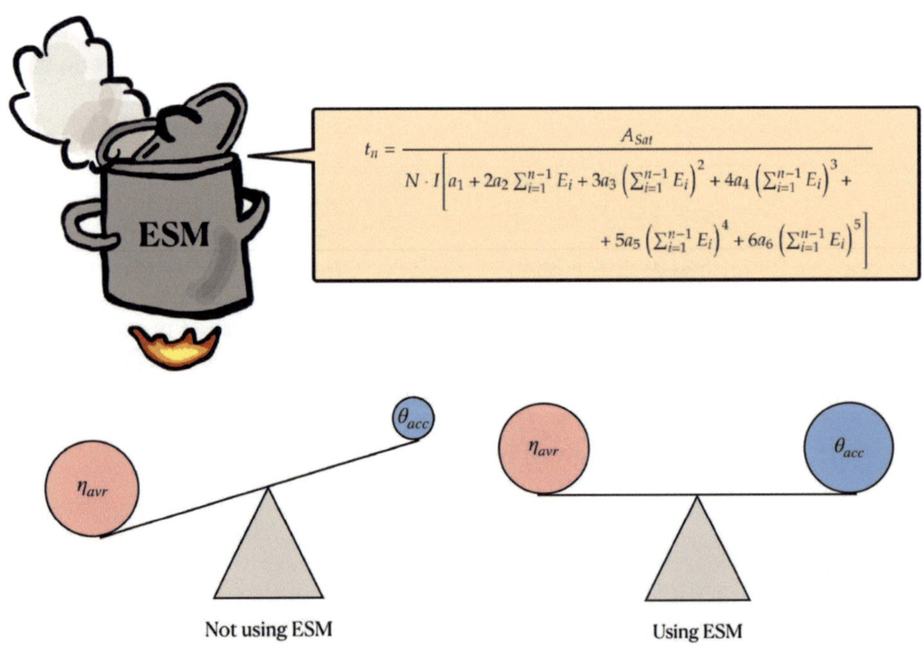

Figure 6.
Illustration showing how ESM is a useful tool to increase the average diffraction efficiency and consequently the acceptance angle of multiplexing-based holographic concentrators.

where A_{Sat} is the experimentally determined value of the saturation grating intensity, N is the total number of holograms recorded up to the saturation point (thus A_{Sat}/N is the grating intensity to record N holograms with the same diffraction efficiency), E_n is the exposure energy density required to record the *nth* hologram of the sequence, and E_i is the amount of exposure energy required to record the previous *ith* holograms of the sequence. Algebraically manipulating this expression and introducing the polynomial function for A_N yields the exposure energy needed to record the *nth* hologram of the sequence with the same efficiency as the others. Finally, the time of each exposure can be obtained from the following equation

$$t_n = \frac{A_{Sat}}{N \cdot I\left[a_1 + 2a_2\sum_{i=1}^{n-1}E_i + 3a_3\left(\sum_{i=1}^{n-1}E_i\right)^2 + 4a_4\left(\sum_{i=1}^{n-1}E_i\right)^3 + +5a_5\left(\sum_{i=1}^{n-1}E_i\right)^4 + 6a_6\left(\sum_{i=1}^{n-1}E_i\right)^5\right]} \quad (15)$$

Figure 6 shows an illustration of how ESM is a useful tool in HSC design.

4. Methods

4.1 Green prepolymer preparation

In the present work, volumetric phase transmission multiplexed HLs have been recorded in a green photopolymer named Biophotopol, which is a low-toxicity material containing one or more monomers in a binder, an electron donor, and a sensitizer dye [7, 18, 19]. The composition of Biophotopol that was used can be seen in **Table 1**. Among the compounds that form it are NaAO (sodium acrylate) as in situ generated monomer, TEA (triethanolamine) as initiator and plasticizer, RF (riboflavin 5′-monophosphate sodium salt) as a dye, water as a solvent, and PVA (polyvinyl alcohol) as a binder (degree of hydrolysis = 87.7%, Mw = 130,000 $g \cdot mol^{-1}$). The prepolymer solution was deposited in $6.5 \times 6.5\ cm^2$ glass molds and allowed to dry in the dark inside an incubator for 24 h at a controlled relative humidity of $60 \pm 5\%$ and a temperature of $20 \pm 1^{\circ}C$. Layers of 100 to 200 μm thickness for holographic recording were prepared from different volumes of prepolymer solution.

4.2 Holographic setup and solar simulator

The experimental setup used for recording consisted of a diode-pumped laser emitting continuously at 473 nm, which was spatially filtered and separated into two

PVA (% w/w)[a]	NaAO (M)[b]	TEA (M)	RF (M)
11.3	0.49	$1.7 \cdot 10^{-3}$	$111 \cdot 10^{-3}$

[a]*w/w: weight by weight.*
[b]*M: molarity.*
PVA (polyvinyl alcohol), NaAO (sodium acrylate), TEA (triethanolamine), and RF (riboflavin 5′-monophosphate sodium salt).

Table 1.
Optimized concentrations of the prepolymeric composition.

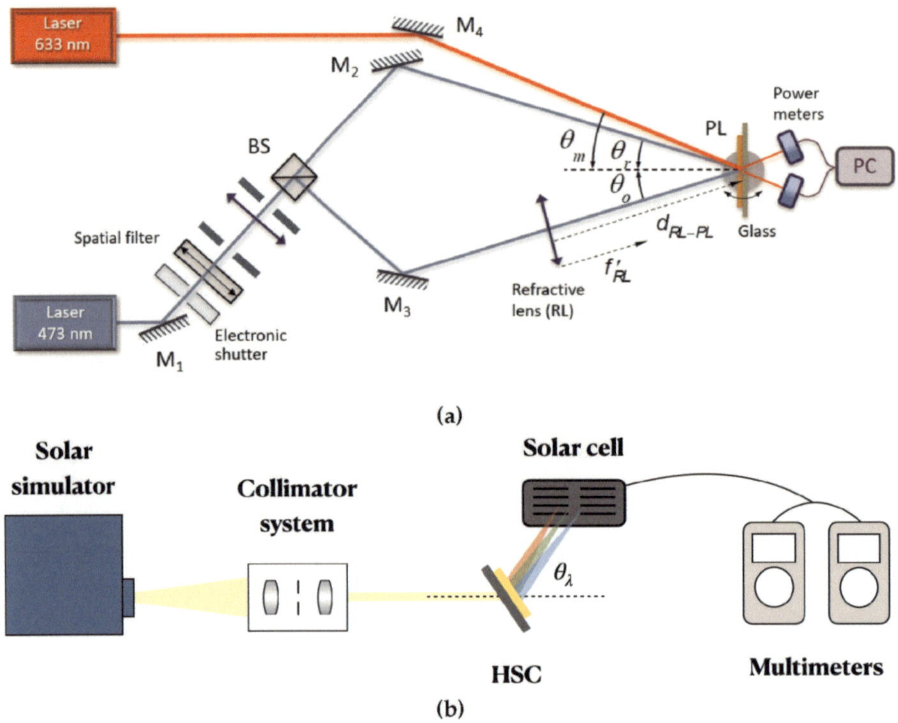

Figure 7.
Experimental setups. (a) Recording of the HLs using a 473 nm laser and reconstruction of the HLs using a He-Ne laser emitting at 633 nm. (b) Reconstruction of holographic solar concentrators using the solar simulator.

secondary beams (object and reference beam) using a 1:1 beam splitter cube. The recording intensity of the setup was 2.2 mW/cm^2. A collimated reference beam and a diverging object beam were directed onto the photopolymer, creating an interference pattern. The divergent beam was generated by using a refractive lens. **Figure 7(a)** shows the experimental setup used during the recording stage. In addition, to optimize the exposures of each of the multiplexed holograms, the recording was monitored using an unexpanded He-Ne laser emitting at a wavelength of 633 nm, to which the dye (RF) of the recording material was not sensitive.

A solar simulator was used to characterize the holographic solar concentrators, the experimental setup of which is shown in **Figure 7(b)**. The HSCs were characterized using a broadband nonpolarized source and a monocrystalline silicon photovoltaic cell (PHYWE 2.5 × 5 cm^2). The relative value of the short-circuit current under short-circuit conditions ($R = 0\ \Omega$) was measured. For this purpose, a standard solar simulator source (model 10,500, ABET Technologies) was used to emit a continuous solar spectrum (350–1800 nm), which was collimated to reconstruct the holographic lenses using the convergent conjugated beam to obtain the real image. The diffracted beam of the holographic lenses focused the light onto the photovoltaic cell as an extended focus. However, the chromatic dispersion of the diffracted beam produced an aberrated image on the surface of the solar cell.

5. Results

In this chapter, we have studied the most important parameters to optimize the multiplexing conditions and thus obtain holographic solar concentrators with high diffraction efficiency and wide acceptance angle [6, 7, 10, 17].

5.1 Influence of the number of HLs

First of all, the influence of the number of multiplexed HLs has been analyzed. **Figure 8(a)** shows that the higher the number of multiplexed HLs, the average diffraction efficiency decreases. **Figure 8(b)** shows that the higher the number of multiplexed HLs, the higher the acceptance angle. The key is to choose the number of HLs that balances both parameters. In this case, the optimal number was five multiplexed HLs.

5.2 Influence of the photopolymer physical thickness

Once the optimal number (five multiplexed HLs) was chosen, the influence of the physical thickness of the photopolymer layer was studied. It is logical that in order to multiplex a larger number of HLs, the thickness of the photopolymer layer has to be thicker than the optimal layer for 1 HL recording. Thicknesses from 100 to 200 μm were studied. **Figure 9** shows the results obtained for different thicknesses. The best results were obtained for thicknesses from 170 to 180 μm.

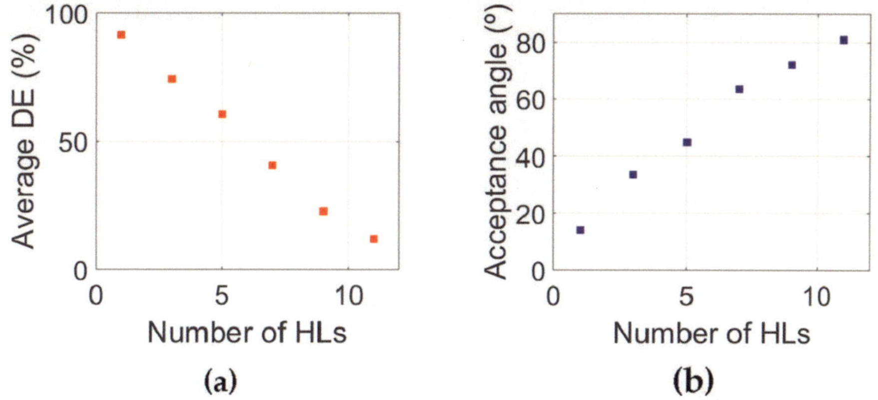

Figure 8.
Influence of the number of multiplexed HLs on (a) average diffraction efficiency, (b) acceptance angle.

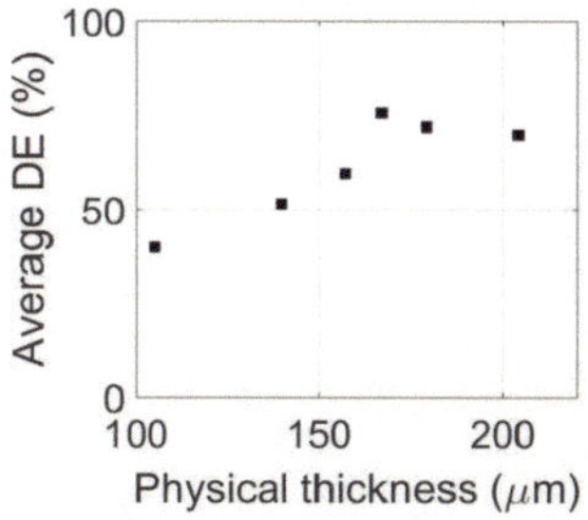

Figure 9.
Influence of photopolymer layer thickness on average diffraction efficiency.

5.3 Influence of peak-to-peak angular distance

The influence of the peak-to-peak angular distance on the multiplexing conditions has also been studied. It has been found that as the peak-to-peak angular distance decreases, the average diffraction efficiency increases (the average diffraction efficiency was 73, 82, and 85%, respectively). However, the acceptance angle decreases (the acceptance angles were 50°, 38°, and 26°, respectively). It is noteworthy that,

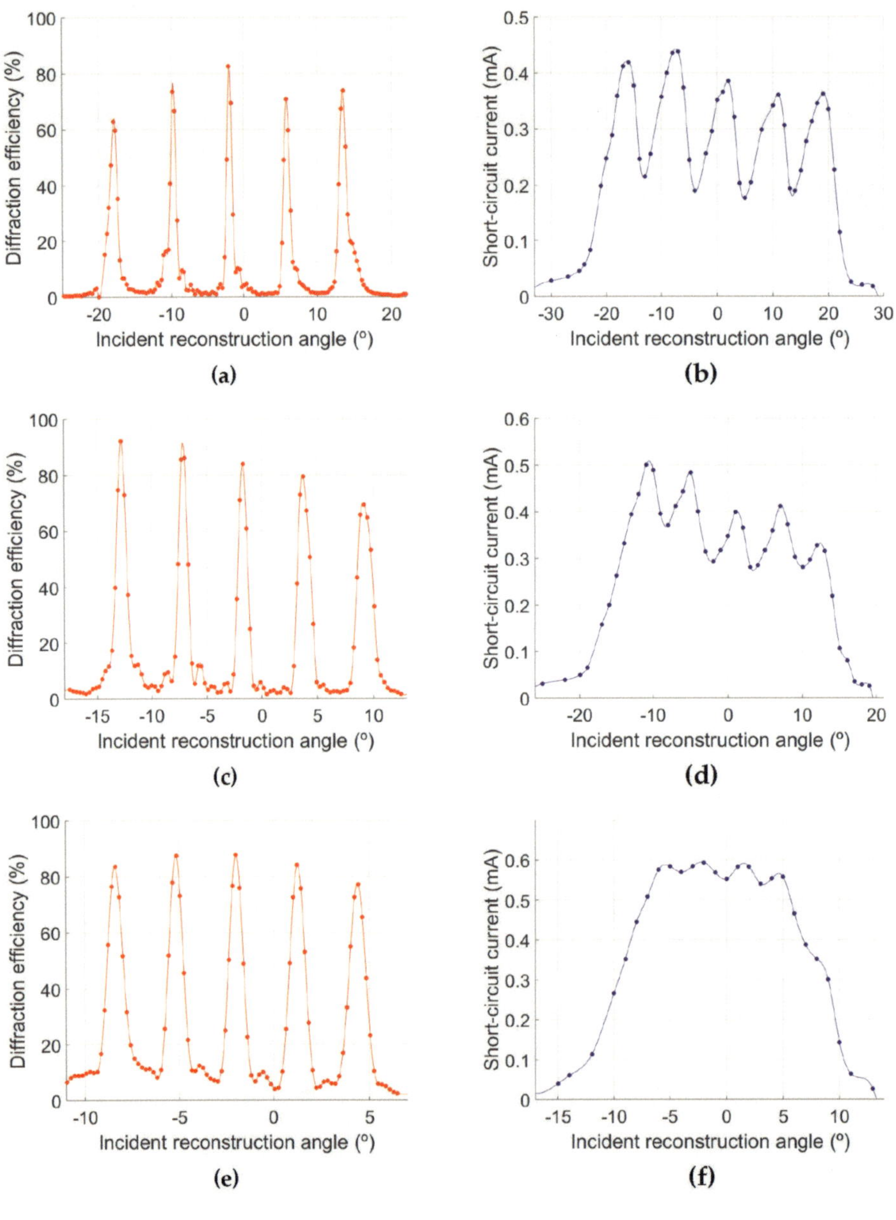

Figure 10.
Influence of peak-to-peak angular distance. (a, b) angular distance of 7.3°, (c, d) angular distance of 5°, and (e, f) angular distance of 3°. Reconstructions with (a, c, e) 633 laser, and (b, d, f) solar simulator.

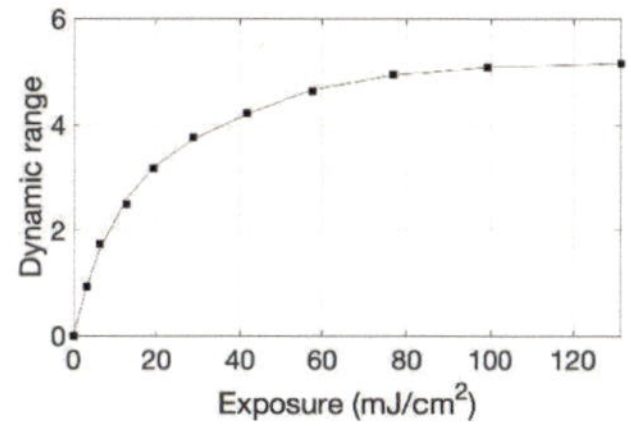

Figure 11.
Dynamic range as a function of exposure energy.

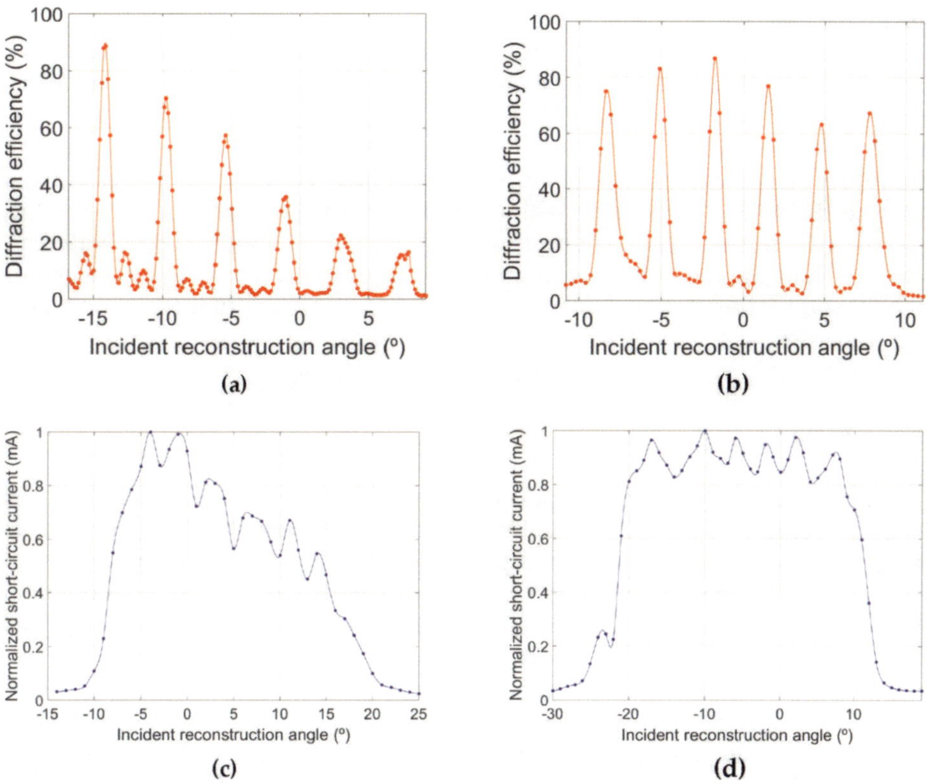

Figure 12.
HSC based on six multiplexed HLs (a, c) not using ESM, (b, d) using ESM. Reconstructions with (a, b) 633 nm laser, (c, d) solar simulator.

although the acceptance angle decreases, the peak-to-peak drop is smaller when this distance is smaller, which is a great advantage since the area under the curve is much more uniform.

Figure 10 shows reconstructions with a He-Ne laser emitting at 633 nm (**Figure 10 (a)**, **(c)**, and **(e)**) and with the solar simulator (**Figures 10(b)**, **10(d)**, and **10(f)**) for holographic solar concentrators based on five multiplexed HLs with different peak-to-peak angular distances. **Figure 10(a)** and **(b)** are HSCs with peak-to-peak angular distance of 7.3°, in **Figure 10(c)** and **(d)**, it was 5°, and finally in **Figure 10(e)** and **(f)**, it was 3°.

5.4 Influence of using ESM

Finally, the potential use of the exposure schedule method in the field of holographic solar concentrators was studied. ESM is a method that has been commonly used in the field of data storage where hundreds of holograms have been multiplexed, so in this work, we tried to adapt this method for a few holograms. First of all, we started by performing the Exposure vs. Dynamic Range ($M\#$) curve in order to obtain the A_{sat} parameter. **Figure 11** shows this curve, which was fitted with a polynomial function defined by Eq. (13) and where $A_{sat} = 5.2$.

Figure 12 shows reconstructions with the 633 nm laser (**Figure 12(a)** and **(c)**) and with the solar simulator (**Figure 12(b)** and **(d)**) for holographic solar concentrators based on six multiplexed HL. **Figure 12(a)** and **(b)** are without using ESM and **Figure 12(c)** and **(d)** are using ESM.

On the one hand, it can be seen that using ESM, the diffraction efficiency of the last peaks is significantly improved compared to not using ESM. It goes from an average diffraction efficiency of 45–75%. On the other hand, if we compare the results obtained to the solar simulator, it can be seen that although the acceptance angle does not improve considerably (from 30° to 35°), if the area under the curve is calculated, it goes from 18.9 to 30.4, which is a really significant improvement.

Despite the fact that a large number of HLs have not been multiplexed with respect to current HSC, the use of ESM is an important step for the design of future generation HSC and for the development of future photopolymers with increasingly less toxic properties.

6. Conclusions

The advancement of new technologies must be aligned with environmental sustainability. Hence, one of the overarching goals is to facilitate the energy transition to clean energy sources such as solar energy. In this chapter, we have focused on recapitulating and studying the most important optical tools used in the development of holographic solar concentrators. The current limitations and perspectives of HSCs lie in the improvement of the photopolymers used as holographic recording materials. There is a trend toward the development of new photopolymers with attractive properties for next-generation holographic devices applied to the solar energy field. The multiplexing technique of HLs enables higher chromatic selectivity, which is essential for the design of potential hybrid systems, as it allows the full solar radiation to be exploited and increases the efficiency of the system. In the design of future photopolymers, it is necessary to focus on the optimization of compositions that allow increasing the dynamic range to multiplex more HLs in the same hologram while maintaining a high diffraction efficiency and thus further increasing the acceptance angle.

Acknowledgements

This research was funded by Universidad de Alicante grant UAFPU2023; MICIU/AEI/10.13039/501100011033 grants PID2019-106601RB-I00, PID2023-148881OB-I00; Generalitat Valenciana grants CIDEX/2022/60, IDIFEDER/2021/014, PROMETEO/2021/006.

Abbreviations

HOE	holographic optical element
HL	holographic lens
RL	refractive lens
HSC	holographic solar concentrator
ESM	exposure schedule method
PVA	polyvinyl alcohol
NaAO	sodium acrylate
TEA	triethanolamine
RF	riboflavin 5′-monophosphate sodium salt

Author details

Tomás Lloret[1*], Marta Morales-Vidal[1,2] and Inmaculada Pascual[1,2]

1 Departamento de Óptica, Farmacología y Anatomía, Universidad de Alicante, San Vicente del Raspeig, Spain

2 Instituto Universitario de Física Aplicada a las Ciencias y las Tecnologías, Universidad de Alicante, San Vicente del Raspeig, Spain

*Address all correspondence to: tomas.lloret@ua.es

References

[1] Xie W, Dai Y, Wang R, Sumathy K. Concentrated solar energy applications using Fresnel lenses: A review. Renewable and Sustainable Energy Reviews. 2011;**15**:2588-2606

[2] Abojela Z, Desa M, Sabry A. Current prospects of building-integrated solar PV systems and the application of bifacial PVs. Frontiers in Energy Research. 2023; **11**:1-23

[3] Ludman J. Holographic solar concentrator. Applied Optics. 1982;**21**: 3057-3058

[4] Sheridan J, Kostuk R, Gil A, Wang Y, Lu W, Zhong H, et al. & others roadmap on holography. Journal of Optics. 2020; **22**:123002

[5] Alfaro E, Lloret T, Vilardy J, Bastidas M, Morales-Vidal M, Pascual I. Photopolymer holographic lenses for solar energy applications: A review. Polymers. 2024;**16**:1-24

[6] Morales-Vidal M, Lloret T, Ramírez MG, Beléndez A, Pascual I. Green and wide acceptance angle solar concentrators. Optics Express. 2022;**30**: 25366-25379

[7] Lloret T, Morales-Vidal M, Nieto-Rodríguez B, García-Vázquez J, Beléndez A, Pascual I. Building-integratedconcentrating photovoltaics based on a low-toxicity photopolymer. Journal of Physics: Energy. 2024;**6**:015017

[8] Naydenova I, Akbari H, Dalton C, Yahya M, Ilyas M, Wei C, et al. Photopolymer holographic optical elements for application in solar energy concentrators. In: Mihaylova E, editor. Holography—Basic Principles and Contemporary Applications. London, United Kingdom: IntechOpen; 2013. pp. 129-145

[9] Kumar A, Deo N, Yadav H. Analysis of design parameters for wavelength selective holographic solar concentrators. In: 2008 33rd IEEE Photovoltaic Specialists Conference. San Diego, CA, USA: IEEE; 2008. pp. 1-4

[10] Lloret T, Morales-Vidal M, Nieto-Rodríguez B, García-Vázquez J, Ramírez MG, Beléndez A, et al. Shrinkage studies and optimization of multiplexed holographic lenses with high diffractive efficiency and wide angular response. Optics and Photonics for Information Processing XVII. 2023; **12673**:105-111

[11] Lloret T, Navarro-Fuster V, Ramírez MG, Ortuño M, Neipp C, Beléndez A, et al. Holographic lenses in an environment-friendly photopolymer. Polymers. 2018;**10**:302

[12] Champagne E. Nonparaxial imaging, magnification, and aberration properties in holography. Journal of the Optical Society of America A. 1967;**57**:51-55

[13] Kostuk R. Holography: Principles and Applications. Boca Raton, FL, USA: CRC Press; 2019

[14] Kogelnik H. Coupled wave theory for thick hologram gratings. The Bell System Technical Journal. 1969;**48**: 2909-2947

[15] Wu H, Park S, Lee J, Jeon S, Kim N. Holographic solar concentrator using multiplexed holographic optical elements. In: 2018 23rd Opto-Electronics and Communications Conference (OECC). Jeju Island, Korea South: IEEE; 2018. pp. 1-2

[16] Fernández E, Ortuño M, Gallego S, García C, Beléndez A, Pascual I.

Comparison of peristrophic multiplexing and a combination of angular and peristrophic holographic multiplexing in a thick PVA/acrylamide photopolymer for data storage. Applied Optics. 2007; **46**:5368-5373

[17] Lloret T, Morales-Vidal M, Alfaro E, García-Vázquez J, Nieto-Rodríguez B, Pascual I. Towards sustainable solar energy: Optimization of a holographic concentrator in a green photopolymer. Photosensitive Materials and their Applications III. 2024;**13015**:130150Q

[18] Ortuño M, Fernández E, Gallego S, Beléndez A, Pascual I. New photopolymer holographic recording material with sustainable design. Optics Express. 2007;**15**:12425-12435

[19] Ortuño M, Gallego S, Márquez A, Neipp C, Pascual I, Beléndez A. Biophotopol: A sustainable photopolymer for holographic data storage applications. Materials. 2012;**5**: 772-783

Chapter 4

Perspective Chapter: Recent Developments in Digital Holography Approach Employing Digital Advanced Micromirror Devices

Muhammad Moin, Mehrunisa Moin, A. Qadoos, Abdul Waheed Anwar and Junaid Hassan

Abstract

This investigation explores digital holography using digital micromirror devices (DMDs), examining theoretical foundations and practical applications. We delve into the principles of diffraction, the structure and operation of DMDs, and their holography-relevant properties. The study discusses various methods for producing holographic displays, addressing challenges, and suggesting solutions. Computational strategies for creating holograms for 3D exhibitions employing DMDs and advanced approaches for holographic image modernization are explored. Current findings noted the future of DMDs in holography, such as fast switching speeds, enhancing angular separation between both reconstructed and incident beams, as well as finer mirror pitch and involvement in superior hologram high quality. Also, investigate the employ of DMDs in digital holographic interferometry for a real-time show of interferometric fringes. For faster modulation speed, an electrooptic modulator was employed in synchronization with the DMD in an amplitude modulation mode to create grayscale patterns with a frame rate ~ 833 Hz with a display time noted of only 1.2 ms instead of 38.4 ms for time multiplexing, gaining a speed up by a factor of 32. Experimental results demonstrate the feasibility of DMDs for this purpose, opening new applications in the field.

Keywords: computer-generated holography, digital micromirror device, digital holography, 3-D holographic display, super pixel method

1. Introduction

Digital holography has emerged as a ground-breaking technique that revolutionizes the way three-dimensional information is captured and reconstructed [1]. This innovative approach leverages advanced electronic sensors and sophisticated computer algorithms to record and process holograms, enabling real-time and highly precise imaging [2]. This capability has profound implications across a multitude of fields, including microscopy, data storage, and display technology. Traditional holography, developed in the mid-twentieth century, involves capturing light interference patterns on photographic plates. The process requires a coherent light source, such as a laser, and meticulous alignment of optical components. The interference pattern, or hologram, is recorded on a photosensitive medium and can be reconstructed by illuminating it with the original reference beam [3]. While this method is powerful, it has significant limitations in terms of speed, flexibility, and real-time capabilities. In contrast, digital holography replaces photographic plates with electronic sensors, such as charge-coupled device (CCD) or complementary metal-oxide-semiconductor (CMOS) cameras [4]. These sensors capture the hologram digitally, and computer algorithms process the data to reconstruct the 3D image. This transition to digital recording and processing has several advantages: digital holography enables the real-time capture and reconstruction of holograms, a feat unattainable with traditional methods. SLMs have been widely employed in optical responses such as wave-front influencing and light focusing. However, a major restriction of liquid-type crystal SLMs is their very low refresh amounts, typically less than 120 Hz which hampers high-speed responses, such as beam focusing in live materials [5]. This has paved the way for DMDs to be used for similar applications due to their much faster refresh rates, with a maximum of 32 kHz [6]. Although DMDs allow only binary amplitude modulation, compared to the grayscale and phase modulation of SLMs, they have been shown to excel in beam-shaping applications. For instance, Conkey et al. [7] investigated the employ of new computer-generated holograms to gain beam focusing in the scattering area with applied a signal-to-background ratio and noted values of up to ~160 in ~34 ms, respectively. Additionally, the DMD as binary data in a reference-less optical material for beam focusing, employing a Bayesian phase repossession algorithm to approximate the transmission matrix and focus beams through a highly scattering medium and binary-based digital optical phase conjugation system using DMDs, achieving speed enhancements by approximately two orders of magnitude compared to liquid crystal SLMs, with a low playback latency of ~5 ms [8]. In display applications, DMDs modulate time-integrated light intensity through sequential pulse-width modulation. Furthermore, in optical diffraction tomography. Lee et al. [9] demonstrated an improved time-multiplexing structured illumination control scheme using DMDs, achieving better performance with 8-bit amplitude modulation compared to binary amplitude modulation, thereby reducing unwanted artifacts. By leveraging the DMD's built-in dynamic range feature and controlling the frame time, we project an 8-bit image in 38.4 ms (256 × 150 μs). The current approach involves dividing the complex field into real as well as imaginary components and employing the DMD's time modulation system to synthesize a complex signal. This approach is established for wave-front shaping by phase conjugating a computed field in free space or after crossing through a scattering medium. This technique paves the way for fast, digital-accuracy imaging, and optical phase conjugation systems without compromising light efficiency. To further increase the frame rate for grayscale modulation, we modulated the intensity of the illuminating beam for each bit plane while keeping the time constant for each bit plane. This reduced the frame time for each grayscale image to 1.2 ms (8 × 150 μs) [8]. The chapter first introduces the basic concept, followed by a methods section detailing the technique's application to phase conjugation, referred to as optoelectronic phase conjugation to distinguish it from conventional digital phase conjugation. The results section is divided into two main

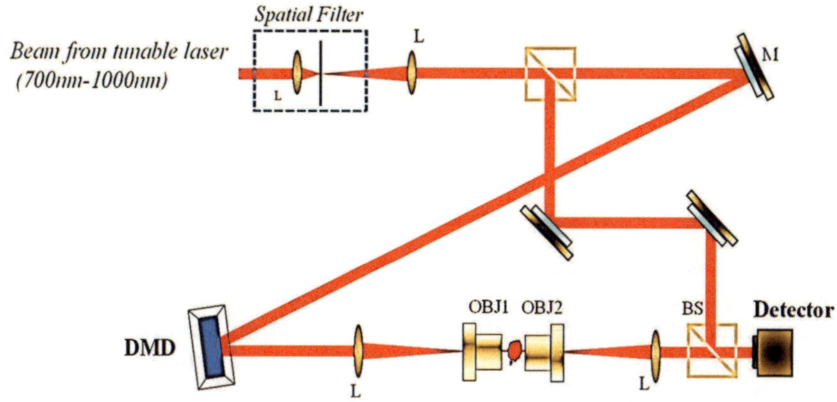

Figure 1.
An optical 4F system is used to image the DMD pattern onto the detector, with the spatial magnification determined by the 4F system configuration" [8].

parts: the first presents experiments demonstrating phase conjugation using our time-multiplexed technique, and the second explores a new modality for DMD use, the amplitude modulation scheme as shown in **Figure 1**.

2. Fundamentals of digital holography

Digital holography is a powerful technique for capturing and reconstructing three-dimensional information using digital means. It has found applications in various fields, such as microscopy, data storage, and display technology [10]. This section explores the basic principles of holography, the distinctions between digital and traditional methods, and the key components and techniques involved in digital holography.

3. Basic principles of holography

Holography, developed by Dennis Gabor in 1948, is based on the principles of interference and diffraction of light waves [11]. It involves recording and reconstructing the light field that is scattered from an object, capturing both its amplitude and phase information. The process can be divided into two main stages: recording and reconstruction.

3.1 Recording stage

In the recording stage, a coherent light source, typically a laser illuminates the object. The light scattered by the object (object beam) interferes with a reference beam, which is derived from the same laser source but does not interact with the object. The resulting interference pattern, or hologram, contains information about both the amplitude and phase of the light waves. This pattern is recorded on a photosensitive medium in traditional holography or an electronic sensor in digital holography.

The mathematical representation of the recorded hologram is given by the interference pattern I(x,y):

$$I(x,y) = \left|O(x,y) + R(x,y)\right|^2 \tag{1}$$

Where $O(x,y)$ is the complex amplitude of the object wave, and $R(x,y)$ is the complex amplitude of the reference wave.

3.2 Reconstruction stage

In the reconstruction stage, the recorded hologram is illuminated with a reference beam, which replicates the original wavefronts that were scattered from the object [12]. This process creates a three-dimensional image of the object that can be viewed from different angles, providing a sense of depth. The reconstructed image is a result of the diffraction and interference of the reference beam with the recorded holographic pattern.

3.3 Classification of digital holography

In digital holography, there are several techniques based on both acquisition and reconstruction methods. In-line digital holography involves collinear objects and reference beams, simplifying the alignment but often leading to overlap between real and virtual images, which can complicate image reconstruction. In contrast, off-axis digital holography separates the beams angularly, making it easier to distinguish between real and virtual images in the Fourier domain. Another approach, phase-shifting digital holography, captures multiple holograms with varied phase shifts in the reference beam, allowing for high-contrast image reconstruction with enhanced phase information. Regarding reconstruction methods, numerical reconstruction

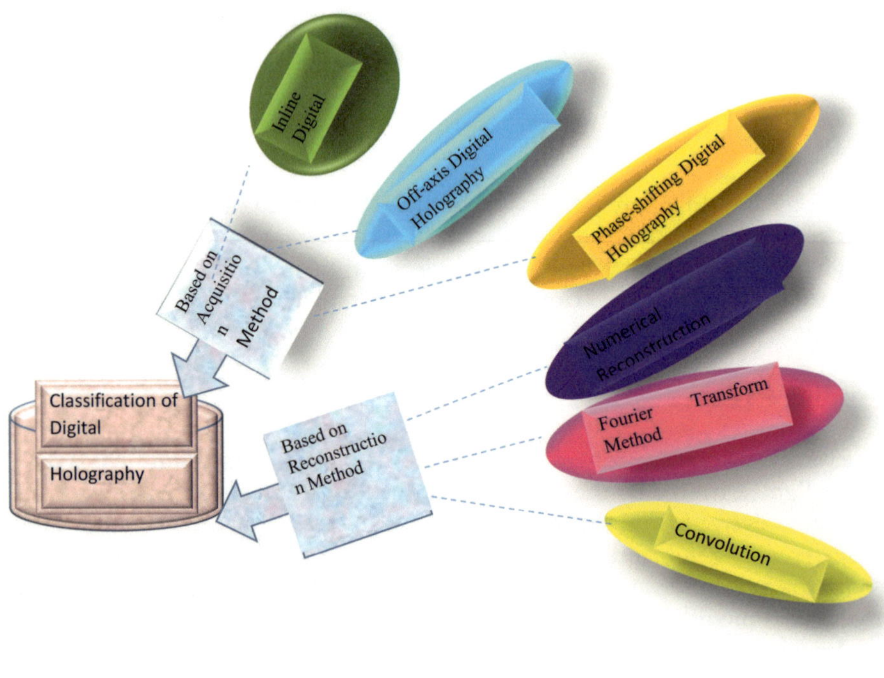

Figure 2.
Classification of advanced holography.

simulates light propagation, offering the advantage of focusing at different depths after the hologram has been recorded. The Fourier transform method operates in the frequency domain and efficiently separates real and virtual images. The convolution method computes the convolution of the hologram with a point-spread function, providing computational efficiency and flexibility. Lastly, the Fresnel approximation is commonly applied in near-field digital holography, using Fresnel integrals to approximate light diffraction. The main classification of digital holography is shown in **Figure 2**.

4. Key components and techniques in digital holography

Digital holography systems comprise several key components and employ various techniques to achieve high-resolution and precise 3D imaging. In digital holography systems, the laser source provides the coherent light necessary for recording holograms digitally. A beam splitter is used to divide this laser beam into two parts: the reference beam and the object beam. The object beam interacts with the object, capturing the scattered light, which contains information about the object's structure. This light is then directed to the digital sensor. Meanwhile, the reference beam does not interact with the object but is instead used to interfere with the object beam, creating an interference pattern that is recorded as the hologram on the sensor. For the recording, a CCD or CMOS camera is employed, replacing traditional holographic plates by capturing the interference pattern digitally. Finally, a computer processing unit takes this recorded hologram and processes it using numerical algorithms, allowing the reconstruction of the object's 3D image. The entire system depends on the precise alignment of the beams and accurate computation for successful hologram reconstruction. The important key components of digital holography as shown in **Figure 3**.

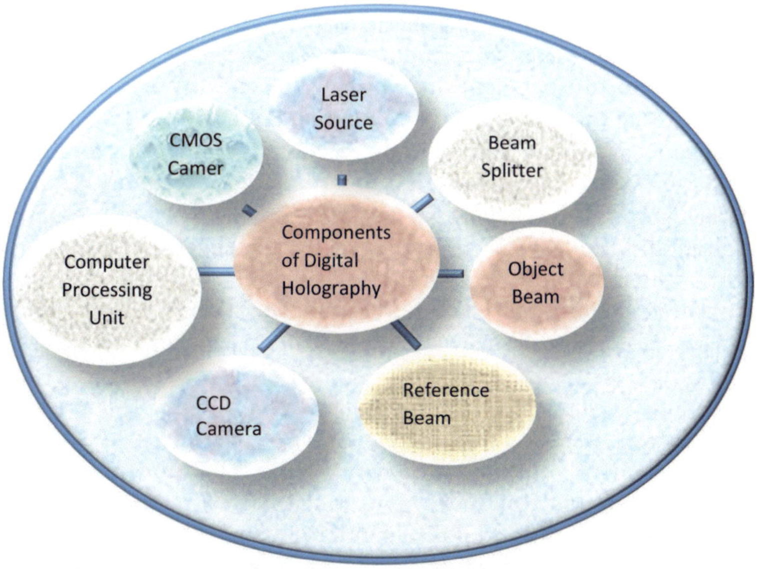

Figure 3.
Key component of advanced digital holography.

4.1 Coherent light source

Lasers are used as coherent light sources due to their ability to produce highly monochromatic and coherent light, which is essential for creating clear interference patterns.

4.2 Beam splitter

A beam splitter divides the laser beam into two parts: the object beam and the reference beam. The object beam illuminates the target, while the reference beam directly reaches the sensor.

4.3 Object and reference beams

The object beam interacts with the object, and the scattered light carries the object's information. The reference beam serves as a stable wavefront for interference.

4.4 Electronic sensor

CCD or CMOS cameras capture the interference pattern. These sensors convert the light pattern into digital data that can be processed by a computer.

4.5 Computer and algorithms

Computational algorithms process the captured interference pattern to reconstruct the three-dimensional image of the object. These algorithms include Fourier transforms, phase retrieval techniques, and digital filtering.

5. Digital micromirror devices (DMDs)

DMDs have significantly impacted the field of digital holography by providing a versatile and efficient means of light modulation. The development of DMD technology is closely linked to advancements in microelectromechanical systems (MEMS). Dr. Larry Hornbeck at Texas Instruments (TI) invented DMDs in 1987 [13]. Initially driven by the need for high-resolution, high-speed light modulators for digital projection systems, DMD technology saw its first functional prototype in 1992, leading to the commercial launch of the first digital light processing (DLP) projector by TI in 1997. Throughout the 2000s, DMD technology expanded beyond projection systems to applications in spectroscopy, 3D scanning, and digital holography, benefiting from continuous improvements in fabrication processes, such as reducing micromirror size and increasing array density. A DMD is composed of an array of microscopic mirrors, each mounted on a tiny hinge that allows individual tilting to direct light toward or away from a target. The micromirrors, measuring a few micrometers across, can tilt ±12 degrees or ±17 degrees, depending on the model. The hinge and yoke assembly enables each micromirror to pivot, offering precise control over light direction through electrostatic actuation generated by electrodes beneath the micromirrors. DMDs operate by controlling the binary state (on/off) of each micromirror. When a micromirror is in the "on" position, it directs light toward the projection lens or target, and in the "off" position, it reflects light away. Rapidly switching the mirrors between these states at rates of up to tens of kilohertz allows DMDs to modulate light effectively, essential for dynamic holographic applications. DMDs provide several advantages in digital holography [1]. Their high-speed capability, with micromirrors switching at rates up to tens of kilohertz, enables real-time holographic displays and dynamic light modulation. Modern DMDs feature arrays with millions of micromirrors, offering the high spatial resolution necessary for detailed holographic patterns. Digital control allows precise modulation of light, while programmability facilitates the generation of complex holographic patterns on the fly, making DMDs highly versatile for applications, such as optical tweezers and 3D imaging. Additionally, DMDs are mechanically robust and have a long operational lifespan, making them suitable for continuous and demanding holography applications. When comparing DMDs with other spatial light modulators SLMs, such as liquid crystal SLMs (LCSLMs) and liquid crystal on silicon (LCoS) SLMs, several distinctions emerge. In terms of speed, DMDs can switch at rates up to tens of kilohertz, making them the fastest among SLMs, while LCSLMs typically switch at rates of a few hundred hertz, and LCoS SLMs offer moderate switching speeds. In terms of resolution, modern DMDs can achieve up to 4 K resolution (3840 × 2160 micromirrors), comparable to or exceeding other SLMs. DMDs provide binary amplitude modulation, generating complex holographic patterns through time multiplexing, whereas LCSLMs and LCoS SLMs offer phase modulation, beneficial for specific holographic applications. Regarding efficiency, DMDs are highly efficient due to their reflective nature and rapid switching, while LCSLMs have lower efficiency due to absorption and scattering within the liquid crystal material, and LCoS SLMs have higher efficiency than LCSLMs but can suffer from polarization dependence. Lastly, in terms of complexity and cost, DMDs are generally less complex and more cost-effective due to their mature manufacturing process, while LCSLMs and LCoS SLMs are more complex and often more expensive. DMDs have been successfully integrated into various digital holography applications, including medical imaging, where they provide high-speed and high-resolution holographic imaging for diagnostics and research. In 3D display systems, DMDs enable real-time holographic displays for entertainment, visualization, and augmented reality. In microscopy, DMDs enhance digital holographic microscopy (DHM) by providing rapid and programmable illumination patterns. Additionally, in optical tweezers, DMDs use holographically generated light patterns to manipulate microscopic particles in biological and physical research. Future research in DMD technology aims to further improve their resolution, speed, and efficiency. Advances in MEMS fabrication techniques and materials science are expected to lead to even more advanced DMDs with smaller micromirrors and faster-switching times. Integrating DMDs with other optical technologies, such as adaptive optics and hybrid SLM systems, could unlock new possibilities for digital holography and beyond [14].

5.1 Recent developments in DMDs

Recent advancements in digital micromirror devices have significantly expanded their capabilities and applications. With increased resolution and pixel density, modern DMDs enable the production of higher-resolution images and offer more precise light modulation, essential for high-definition displays, and scientific imaging. The faster-switching speeds of DMDs allows for real-time data processing, which is critical in fields, such as dynamic holography, medical imaging, and LiDAR systems. Additionally, DMDs have become more versatile due to their ability to operate across a wider range of wavelengths, including ultraviolet and infrared, benefiting applications such as spectroscopy, optical communication, and hyperspectral imaging. Enhanced mirror alignment and control algorithms have improved contrast ratios,

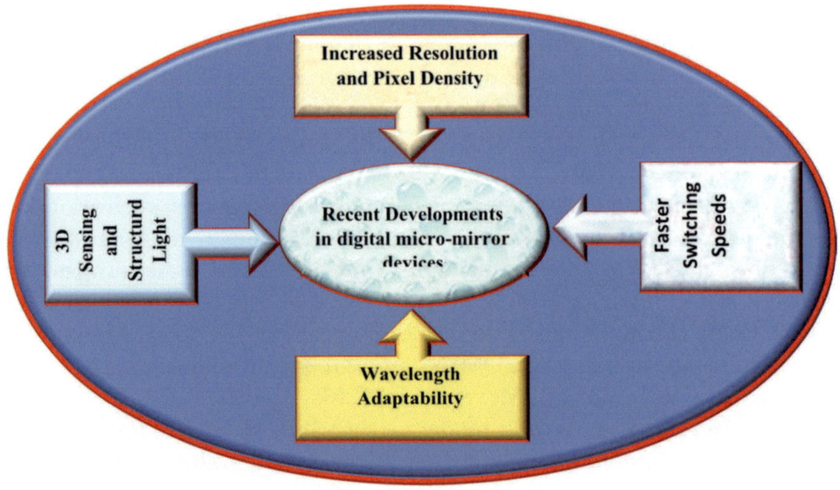

Figure 4.
Recent advancements in digital micromirror devices.

making DMD more effective in high-contrast applications, such as cinema-quality projection and scientific imaging. Furthermore, DMDs are increasingly used in 3D sensing and depth-mapping technologies, where they project structured light patterns to capture 3D data, enabling accurate depth perception in diverse environments. Recent advancements in digital micromirror devices as shown in **Figure 4**.

5.2 Key components in a DMD system

A digital micromirror device system relies on several key components to function effectively. At its core is the micromirror array, which consists of thousands to

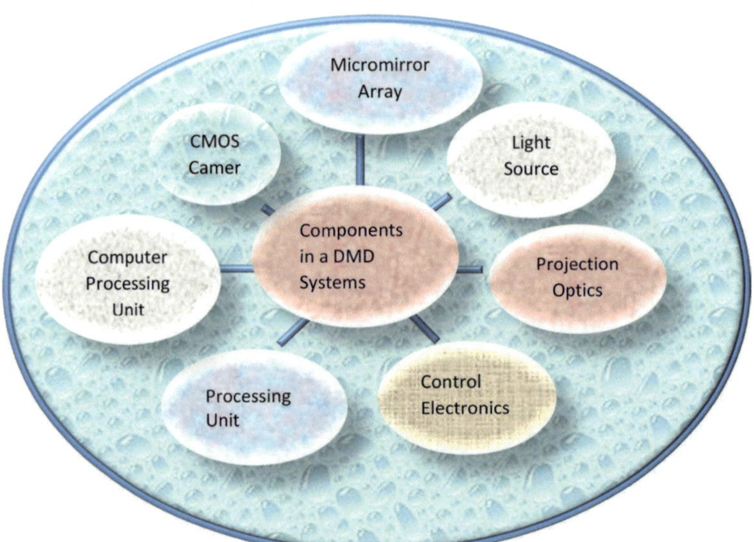

Figure 5.
Key components of digital micromirror.

millions of tiny mirrors that tilt to modulate light in specific directions. The light source, typically a laser or high-intensity LED, provides the initial light that the micromirrors reflect. This modulated light is then directed by projection optics, which focus the light onto a target surface, creating the desired image or pattern. Control electronics play a crucial role by coordinating the movement of the micromirrors and adjusting their tilt to manage the intensity and direction of the light. A processing unit interprets incoming data and generates the necessary control signals for the micromirrors, enabling real-time applications, such as 3D scanning or adaptive optics. Finally, the modulated light is received by either a sensor, for capturing data, or a display surface, for projecting images, completing the system's functionality. The main key component of digital micromirror is shown in **Figure 5**.

6. Recent technological advances in digital micromirror devices (DMDs)

DMDs have undergone significant technological advancements, enhanced their performance, and broadened their applications in digital holography. Texas Instruments (TI) has been at the forefront of these innovations, continuously pushing the boundaries of DMD capabilities. Early DMDs, introduced in the 1990s, had relatively large micromirrors and high switching speeds [15]. Modern DMDs, however, feature micromirrors reduced in size to a few micrometers, increasing the resolution of the DMD array and allowing for more detailed and precise light modulation. This miniaturization results in higher pixel densities, essential for high-resolution digital holography applications. Furthermore, the development of faster electrostatic actuators has enabled micromirrors to switch states at rates exceeding tens of kilohertz, facilitating rapid modulation of light necessary for dynamic holographic displays and real-time imaging systems. The advancements in DMD technology have significantly enhanced their resolution and efficiency. Modern DMDs boast arrays with millions of micromirrors, achieving resolutions up to 4 K (3840 × 2160 pixels) and beyond, which is crucial for creating intricate holographic patterns and detailed 3D reconstructions. High-resolution capabilities are particularly beneficial in digital holographic microscopy, where visualizing fine structural details is essential. Research efforts are also focused on developing DMDs with subwavelength pixel sizes, enabling even finer control over light modulation and pushing the boundaries of image resolution and quality. Efficiency improvements have been another critical area of advancement. The application of advanced reflective coatings on micromirrors has significantly increased their optical efficiency. High-efficiency coatings ensure maximum utilization of incident light, reducing losses and enhancing the brightness and clarity of holographic images. Additionally, innovations in micromirror design, such as optimized hinge and yoke structures, have improved mechanical performance and reliability, reducing mechanical stress and wear and extending the operational lifespan of DMDs. The integration of DMDs with other components in digital holography systems has opened new possibilities for advanced applications. By combining DMDs with adaptive optics systems, real-time correction of optical aberrations becomes possible, which is particularly useful in astronomical imaging and ophthalmology. The use of coherent laser sources with DMDs enables the generation of high-quality holograms by precisely controlling the phase and amplitude of the laser light, facilitating applications in optical trapping, beam shaping, and 3D display technologies. Hybrid systems that combine DMDs with other spatial light modulators, such as liquid crystal on silicon (LCoS) SLMs, offer unique advantages by leveraging the strengths of both technologies. While DMDs provide high-speed binary amplitude modulation, LCoS SLMs offer high-resolution phase modulation, leading to enhanced performance for specific applications. Advances in computational algorithms have also enabled the integration of DMDs with powerful image processing and hologram generation techniques, optimizing the holographic patterns displayed by DMDs and improving the quality and accuracy of holographic reconstructions. This integration is critical for applications in microscopy and 3D imaging, where computational methods enhance hardware capabilities. The technological advancements in DMD design, resolution, efficiency, and integration have significantly expanded the range of applications for DMD-based digital holography. In medical imaging, DMDs enable high-speed and high-resolution holographic imaging systems that provide detailed views of biological tissues and cells, essential for diagnostics and research. In 3D display systems, DMDs facilitate the creation of real-time holographic displays used in entertainment, visualization, and augmented reality, generating dynamic holographic patterns at high speeds for realistic and immersive 3D experiences. In digital holographic microscopy, DMDs enhance imaging processes by providing rapid and programmable illumination patterns, allowing researchers to capture detailed images of microscopic samples with high accuracy. Optical tweezers also benefit from the precise light modulation capabilities of DMDs, using holographically controlled light patterns to manipulate microscopic particles and biological specimens with high precision, aiding research in cellular biology and materials science.

7. Detailed descriptions of recent studies utilizing DMDs in digital holography

In 2019, Lee et al. [16] demonstrated the use of DMDs in high-speed holographic microscopy. The researchers developed a digital holographic microscopy system incorporating a DMD to achieve rapid modulation of light patterns, enabling real-time imaging of live cells. The study highlighted the advantages of using DMDs for dynamic biological imaging, showcasing their capability to capture cellular processes with high temporal resolution. Another significant study in 2005 by Yoon et al. [17] explored the application of DMDs in digital holographic microscopy for noninvasive imaging of living cells. The researchers utilized a DMD to create programmable illumination patterns, enhancing the contrast and resolution of the captured images. This study demonstrated the potential of DMD-based systems in biomedical imaging, providing detailed visualizations of cellular structures without the need for staining or labeling. In the field of optical metrology, conducted experiments using DMDs in digital holographic interferometry. The study focused on the precise measurement of surface topography and deformation using DMD-generated reference wavefronts. The results showcased the accuracy and efficiency of DMD-based interferometry systems, highlighting their application in quality control and materials science.

8. Experimental setups and methodologies

The experimental setups and methodologies in these studies vary depending on the specific application and objectives. However, common elements include the integration of DMDs with coherent light sources, imaging sensors, and computational units for hologram generation and reconstruction. The experimental setup consisted of a DMD illuminated by a coherent laser source [18]. The reflected light from the DMD created dynamic illumination patterns, which were directed onto the biological samples. The resulting holograms were captured by a high-speed camera and processed using advanced algorithms to reconstruct real-time images of the cells. The methodology involved optimizing the DMD patterns to achieve the desired resolution and contrast, allowing for detailed observation of cellular dynamics. The DMD was positioned in the optical path to create programmable light patterns, enhancing the imaging contrast. The captured holograms were processed using phase-shifting algorithms to reconstruct high-resolution images of living cells. The methodology focused on minimizing noise and optimizing the phase shift to achieve accurate and detailed visualizations. The DMD was employed to generate reference wavefronts for interferometric measurements. The experimental setup included a laser source, beam splitters, and imaging sensors to capture the interference patterns. The DMD patterns were programmed to create precise reference wavefronts that interfered with the wavefronts reflected from the measured surfaces. The methodology involved analyzing the interference patterns to extract information about surface topography and deformation with high precision. Advanced investigations used a complex field [19].

$$f(x,y) = A(x,y)e^{j\varphi(x,y)} \tag{2}$$

Where $A(x, y)$ is called the 2D amplitude and $\phi(x, y)$ is represented the 2D phase pattern. We denote a complex pattern on the DMD by showing both positive as well as negative portions of real and imaginary parts of the complex field, sequentially. The grayscale of the real and imaginary parts of the complex field can be shown as follows:

$$f(x,y) = \mathrm{Re}\{|f(x,y)|\} + i * Im\{f(x,y)\} \tag{3}$$

$$\sum_{a=1}^{8} BR_a 2^{a-1} + i * \sum_{a=1}^{8} BI_a 2^{a-1} \tag{4}$$

Phase modulation using DMD as expressed in Eq. (2) by splitting the complex field into real and imaginary parts.

$$f(x,y) = \left[\sum_{a=1}^{8} BR_a^P 2^{a-1} - \sum_{a=1}^{8} BR_a^n 2^{a-1}\right] + i * \left[\sum_{a=1}^{8} BI_a^P 2^{a-1} - \sum_{a=1}^{8} BI_a^n 2^{a-1}\right] \tag{5}$$

Where BR_a^P and BR_a^n are the bit-planes for the both positive and negative portions of the real component, and BI_a^P and BI_a^n are the bit-planes for the positive and negative portions of the imaginary component.

$$U_{o=} T_{io} U_i \tag{6}$$

Where T_{i^o} is the transmission matrix defining the propagation of U_i through the scattering medium.

$$U_{final=} T_{oi} U_o^* \tag{7}$$

$$U_{final} = T_{oi}\left(T_{io} U_i\right)^* = T_{oi} T_{io}^* U_i^* \tag{8}$$

$$U_{final} = T_{oi} T_{io}^* U_i^* = T_{oi} T_{oi}^{-1} U_i^* = U_i^* \tag{9}$$

9. Results and discussions

The results from these studies highlight the significant advantages of using DMDs in digital holography, demonstrating their potential across various applications. The high-speed holographic microscopy system enabled real-time imaging of live cells, capturing dynamic processes, such as cell division and motility. The use of DMDs allowed for rapid modulation of illumination patterns, achieving high temporal resolution and providing detailed insights into cellular behavior. The results underscored the importance of DMDs in dynamic biological imaging, offering a powerful tool for studying live cell dynamics. Researchers achieved noninvasive imaging of living cells with subwavelength axial accuracy using a DMD-based digital holographic microscopy system. The programmable illumination patterns enhanced the imaging contrast, allowing for the visualization of cellular structures without staining. The study demonstrated the potential of DMDs in biomedical imaging, providing a noninvasive and high-resolution technique for observing cellular morphology and dynamics. In 2005, Kreis et al. [20] showed the accuracy and efficiency of DMD-based digital holographic interferometry in optical metrology. The experiments demonstrated precise measurements of surface topography and deformation, highlighting the application of DMDs in quality control and materials science. The results confirmed the capability of DMDs to generate accurate reference wavefronts, enabling high-precision interferometric measurements. These studies collectively emphasize the versatility and effectiveness of DMDs in digital holography. The high resolution, fast switching speeds, and programmability of DMDs make them suitable for a wide range of applications, from dynamic biological imaging to precise optical metrology. The ability to create programmable light patterns and capture high-resolution holograms enables detailed visualization and measurement in various scientific and industrial fields. The digital nature of the data allows for easy manipulation, filtering, and

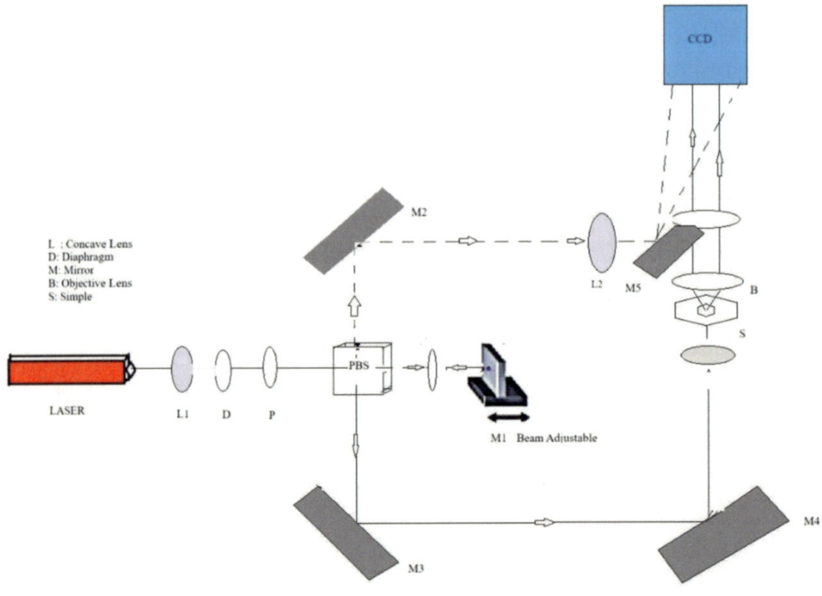

Figure 6.
Schematic of the off-axis digital-holographic microscope setup. Insight, the custom-designed objective; right, aspheric lens system.

enhancement using computational techniques. High-resolution sensors and advanced algorithms offer superior accuracy in capturing and reconstructing fine details. As mentioned earlier, decreasing the wavelength enhances resolution. However, it is crucial to use a practical short wavelength with readily available optical components and detectors. An arrangement operating in the UV spectral range at 193 nm [21]. At this wavelength, the setup does not require a vacuum environment. Optical elements made of fused silica or quartz, which are transparent at 3 nm and commercially available, should be used. **Figure 6** presents a schematic of the setup.

To achieve high resolution, oblique illumination can be employed, shifting the zero-order component to one side of the objective. At the heart of modern digital holography systems is the DMD, a pivotal technology developed by Texas Instruments. DMDs are a type of microelectromechanical system (MEMS) that consists of an array of microscopic mirrors, each capable of independent tilt to modulate light. Originally designed for digital light processing DLP projectors, DMDs have proven to be a game-changer in digital holography due to their speed, precision, and programmability. DMDs can switch mirrors at rates up to tens of thousands of times per second, enabling rapid updating of holographic patterns. This high-speed modulation is critical for applications requiring dynamic holographic displays and real-time image reconstruction. The individual control of micromirrors allows for the creation of highly detailed and precise holographic images. Advances in DMD technology have led to increased mirror density, enhancing the resolution and quality of holograms. DMDs can be programmed to display various light patterns, offering versatility in holographic applications. This programmability allows for adaptive and flexible holographic systems that can be tailored to specific needs. The unique characteristics of DMDs have enabled their application in various fields: In holographic microscopy, DMDs facilitate high-resolution imaging of biological specimens, allowing researchers to observe cellular structures and dynamic processes with unprecedented clarity. Wolfgang Osten et al. [22] also investigated the instance, DMD-based holographic microscopes have been used to study the intricate details of cellular behavior in real-time, providing valuable insights into cellular biology and pathology. **Figure 3** shows a schematic of an opposed-view dark-field digital DHM. The system consists of a symmetric combination of two off-axis dark-field DHMs. Two Nikon bright–/dark-field microscope objectives (MOs) with 20× magnification and a numerical aperture (NA) of 0.45 are positioned face-to-face to capture images from both views. A camera is installed in each view to simultaneously record the dark-field images. Setup of an opposed-view dark-field digital-holographic microscope as shown in **Figure 7**.

DMDs are integral to holographic data storage systems, which utilize the three-dimensional nature of holography to achieve high-density data storage. This technology offers significant advantages over traditional two-dimensional storage methods, including higher capacity and faster data retrieval speeds. The ability of DMDs to rapidly update light patterns makes them ideal for real-time holographic displays used in virtual reality (VR) and augmented reality (AR) applications. These displays provide immersive experiences by generating high quality, interactive 3D images. Recent advancements in DMD technology have further expanded its capabilities and applications in digital holography. Improvements in micromirror fabrication have led to higher mirror densities, enabling the creation of high-resolution holographic images. This increase in resolution is essential for applications requiring detailed and accurate imaging, such as medical diagnostics and material science. New materials and coatings have improved the reflectivity of DMDs, resulting in brighter and more vivid holographic displays. This enhancement is particularly beneficial for applications in

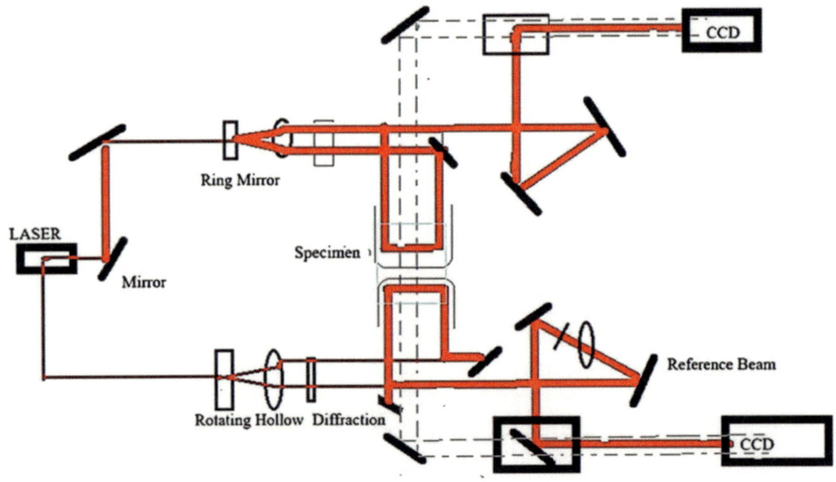

Figure 7.
Setup of an opposed-view dark-field digital-holographic microscope [23].

display technology and scientific visualization. Innovations in driver electronics have reduced latency and increased the speed at which DMDs can switch mirrors. This improvement allows for more complex and dynamic holographic displays, essential for real-time applications in VR and AR [23–25]. Despite the significant progress, several challenges remain in the development and application of DMDs in digital holography. While DMD resolution has improved, further advancements are needed to match the performance of other high-resolution SLMs. The physical size of micromirrors limits the maximum achievable resolution, posing a challenge for applications requiring ultrahigh precision. The high-intensity light used in holographic systems can generate significant heat, affecting the performance and lifespan of DMDs. Effective thermal management solutions are crucial to ensuring reliable operation in high-power applications. DMDs can have high-power requirements, especially in systems that operate continuously or at high speeds. Reducing power consumption is essential for making DMD-based systems more energy-efficient and sustainable. Looking forward, the integration of AI and ML algorithms with DMD-based holography systems promises to enhance image processing, reconstruction quality, and real-time capabilities. AI-driven algorithms can optimize hologram generation and improve the interpretation of complex holographic data, opening new avenues for applications in medical diagnostics, industrial inspection, and beyond [26]. Digital holography, empowered by the advanced capabilities of digital microscope devices, is a transformative technology with far-reaching implications. The high-speed modulation, precision, and programmability of DMDs have revolutionized various fields, from high-resolution microscopy to dynamic holographic displays. As technological advancements continue to push the boundaries of what is possible, the future of digital holography looks promising, with potential breakthroughs in AI integration, wearable devices, and miniaturization on the horizon. The continued research and development in this area will undoubtedly lead to even more innovative applications and enhanced capabilities, solidifying the role of digital holography as a cornerstone technology in the twenty-first century. In 2012, introduced a robust iterative optimization technique known as the genetic algorithm [25]. Its strong robustness has led to widespread application and development in subsequent wavefront-shaping research.

Wu et al. [27] studied the algorithm's flowchart and compared the unoptimized speckle field to the focused light field post-optimization using the genetic algorithm in **Figure 8(a–c)**. The results showed that the light effectively focused on a single focal point after optimization. In 2020, Luo et al. [24] proposed a hybrid approach that integrates the genetic algorithm with deep neural networks, significantly enhancing the speed and performance of wavefront shaping. In 2021, Zhao et al. [29] combined the genetic algorithm with the bat algorithm, achieving high robustness and performance without requiring parameter adjustments. Besides integrating the genetic algorithm with other intelligent optimization techniques, Feng et al. [30] proposed an improved fitness function, enhancing the optimization's effectiveness through a novel discriminant. In summary, recent advancements in wavefront shaping research have continually refined the genetic algorithm. Light successfully focused to a single point after optimization using a genetic algorithm as shown in **Figure 8(a–c)**.

Figure 9(a and **b)** shows the experimental approach and basic principle of OPC. The propagation direction of the light wave is effectively reversed, giving the impression of time reversal while maintaining identical phases at each point. However, since time reversal is not physically feasible, researchers proposed keeping the time component unchanged and instead reversing the sign of the spatial phase term in E_0. The phase conjugate mirror effect can be realized through a combination of a camera and a spatial light modulator, enabling the backward propagation of the light wave while maintaining the integrity of the wavefronts. The digital OPC experimental setup is shown in **Figure 5**. **Table 1** provides a comparative overview of the parameters for DM, LC-SLM, and DMD.

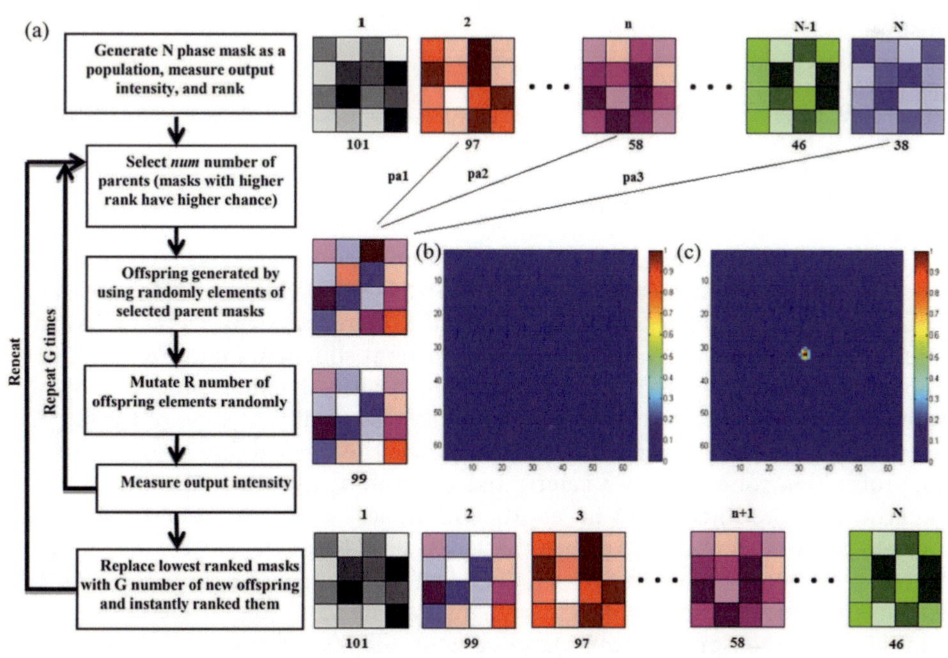

Figure 8.
(a) Genetic algorithm flowchart, (b) without optimized speckle field, (c) light focused to a single point after optimization employing genetic algorithm [28].

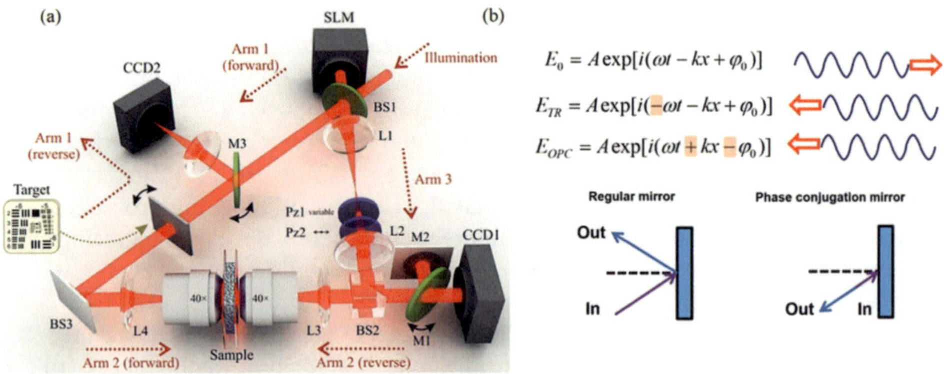

Figure 9.
(a) Digital OPC experimental setup; (b) basic principles of OPC [31].

Types of wavefront modulators	Technology type	Resolution	Adjustable range	Refreshing rate	Enhancement factor α	Capable of withstanding pulsed laser
Deformable mirror (DM)	Deformable mirror	10–1000 pixels	Surface shape modulation	1–10 kHz	π/4	Not specified
Liquid crystal spatial light	Liquid crystal	1920 × 1080 pixels	Phase or intensity modulation	10–100 Hz	π/4	40 mJ/cm^2
Digital micromirror device (DMD)	Micromirror array	3000 × 2000 pixels	Binary on/off modulation	Up to 23 kHz	1/2π	200 μJ/cm^2

Table 1.
A comparative analysis of the parameters for representative wavefront modulators [32].

10. Image plane digital holography

To describe the experimental setup in this context, the system is based on a typical in-line phase-difference holography (IPDH) setup, where a wedge separates a laser beam into two paths: an object beam and a reference beam. These paths are directed using mirrors in two optical guiding arms as shown in **Figure 10(a)**. The stronger of the two beams, referred to as the illumination beam, is processed by several optical elements. Spherical lenses focus the beam at the center of the target object, while a cylindrical lens extends the beam in the vertical direction, creating a slim profile along one axis and an elongated shape along the other. The illumination beam traverses the object along its length, and the scattered light from the object is captured as the object beam. An objective lens is used to image the object near the plane of the camera sensor. In parallel, the reference beam passes through a lens that generates a divergent wavefront, with a virtual source positioned at the same distance from the camera sensor as the objective's aperture. This configuration allows the system to function as a lensless Fourier transform holography setup. This arrangement has the advantage of easily separating the real image from the virtual image in the Fourier domain of the reconstructed hologram, a crucial feature for improving

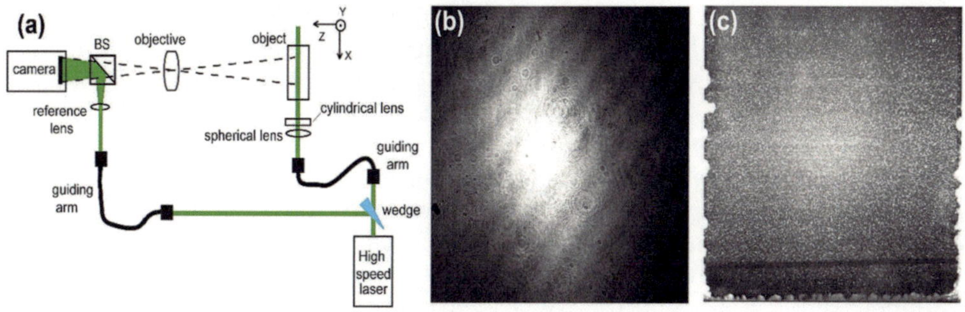

Figure 10.
(a) Schematic drawing of the experimental setup; representative images of the (b) reference beam and (c) object beam.

the performance of the technique. The laser beam, with a diameter of 1.5 mm, can be treated as an extended light source. For optimal results, the expansion rates and orientations of both the object and reference beams need to be well aligned at the sensor to maintain spatial coherence. Since the laser is linearly polarized, both beams were aligned along the same polarization axis to ensure coherence. Specifically, the guiding arms were setup so that both beams were polarized vertically, along the Y-axis in **Figure 10(a)**. In the subsequent analysis **Figure 10(b)** and **(c)**, the object beam and the conjugate (denoted as *) are presented. To highlight the relevant spatial information, all 2D black-and-white images throughout the paper are rescaled such that the top 5% of the brightest pixels are set to the maximum intensity value (255), enhancing contrast and clarity. Schematic drawing of the experimental setup imaging of references as shown in **Figure 10(a–c)**.

11. Challenges and limitations in digital holography using DMDs

DMDs have significantly advanced the field of digital holography, but there are still several challenges and limitations that need to be addressed to fully realize their potential. One of the primary challenges in digital holography using DMDs is the presence of optical aberrations and alignment issues. Precise alignment of optical components is critical to ensuring the accuracy of the holographic reconstruction. In 1962, Leith et al. suggested that misalignments can lead to distortions and reduced image quality, impacting the reliability of measurements and visualizations [33]. Additionally, speckle noise, caused by the interference of coherent light, is a significant challenge in digital holography. This granular noise can degrade image quality and obscure fine details, making it difficult to obtain accurate and high-resolution holograms. In 1976, Kumar et al. [21] found that speckle reduction techniques are necessary to improve the clarity and fidelity of holographic images. Another substantial challenge is the computational load required for processing holographic data. The generation and reconstruction of holograms involve complex algorithms that can be computationally intensive. This challenge is particularly relevant for real-time applications, where rapid processing is essential. Moreover, digital holography systems are often sensitive to environmental factors, such as vibrations, temperature fluctuations, and air currents. In 2005, Kreis et al. [34] determined the factors that can introduce noise and artifacts into the holographic recordings, affecting the accuracy and stability of the measurements. Despite their many advantages, DMD technology has inherent limitations. Although DMDs offer high resolution, their pixel size and density are still limited by current manufacturing technologies. This limitation constrains the achievable resolution of holograms, particularly for applications requiring extremely fine details. In 2005, Lee et al. [35] estimated that DMDs typically operate in a binary mode, where each micromirror is either fully on or off. This binary modulation can limit the grayscale representation and dynamic range of the holographic images, affecting their quality and contrast. Advanced techniques, such as pulse-width modulation, are used to mitigate this limitation, but they can introduce additional complexity. Furthermore, the mechanical nature of DMDs, involving the continuous tilting of micromirrors, can lead to wear and tear over time. This mechanical wear can impact the reliability and longevity of DMD-based systems, especially in applications requiring continuous operation. DMDs can also generate significant heat during operation, especially when handling high-intensity light sources. Effective heat dissipation mechanisms are essential to maintain the performance and longevity of the DMDs, adding to the design complexity of the holographic systems. Advanced alignment and calibration techniques are being explored to address optical aberrations and misalignments in holographic systems. Automated alignment systems and adaptive optics can dynamically correct misalignments, improving the accuracy and reliability of these systems [36]. Speckle noise reduction can be achieved through multiwavelength illumination, spatial filtering, and digital image processing algorithms. Efficient algorithms for hologram generation and reconstruction are crucial, aided by parallel computing, GPU acceleration, and machine learning, which can significantly reduce processing times and enable real-time holography applications [37]. Environmental control systems and innovations in DMD design can help mitigate the sensitivity of holographic systems to environmental factors. Smaller pixel sizes, higher pixel densities, and improved micromirror materials can enhance resolution and reliability. Alternative modulation schemes, such as analog or multilevel modulation, can improve the dynamic range and quality of images. Integrating DMDs with other SLMs, such as LCoS devices, can provide superior performance. Advanced heat management solutions, such as improved thermal interface materials and active cooling systems, can address heat dissipation challenges in DMD-based systems [38].

12. Summary of key points

Digital holography leverages electronic sensors and sophisticated computer algorithms to record and process holograms, providing real-time and precise three-dimensional imaging capabilities. The incorporation of DMDs, originally developed for DLP projectors, has significantly enhanced the speed, precision, and programmability of holographic systems. Despite these advancements, challenges such as optical aberrations, speckle noise, computational load, and environmental sensitivity persist. Limitations specific to DMD technology, including resolution constraints, binary light modulation, mechanical wear, and heat dissipation, also pose hurdles.

13. The significance of recent advancements

Recent technological innovations have made substantial contributions to overcoming these challenges. Advanced alignment and calibration techniques, speckle reduction methods, and more efficient computational algorithms have improved the accuracy, clarity, and processing speed of digital holography systems. Innovations in DMD design, such as developing smaller pixel sizes, higher densities, and improved materials, have enhanced resolution and reliability. Hybrid systems combining DMDs with other spatial light modulators and advanced heat management solutions have further expanded the capabilities and applications of digital holography. These advancements have not only improved the performance of existing applications but have also enabled new possibilities in fields, such as dynamic biological imaging, noninvasive medical diagnostics, and precise optical metrology. The ability to capture high-resolution and real-time holographic images has opened up new avenues for research and practical applications, making digital holography a versatile and powerful tool [39].

14. Final thoughts on the future of digital holography with DMDs

The future of digital holography with DMDs is exceedingly promising. As technology continues to evolve, further improvements in DMD design and integration are expected to address current limitations and expand the range of applications. The ongoing development of more advanced computational techniques and environmental control systems will enhance the stability, accuracy, and efficiency of holographic systems. Moreover, the interdisciplinary nature of digital holography, combining optics, electronics, and computer science suggests fertile ground for future innovations. The integration of artificial intelligence and machine learning with digital holography could lead to new breakthroughs in image processing and analysis, further expanding its applications in science, medicine, and industry. The advancements in digital holography with DMDs have already transformed the field, providing high-resolution and real-time imaging capabilities that were previously unattainable. Continued research and development will likely yield even more significant breakthroughs, solidifying the role of digital holography as a critical technology in the future of imaging and measurement. The potential for DMD-based digital holography to revolutionize various fields is immense, promising a future where precise, real-time, and high-resolution three-dimensional imaging becomes ubiquitous [8].

15. Applications of DMDs in digital holography

DMDs have become indispensable in various digital holography applications due to their high-speed and high-resolution light modulation capabilities. This section explores the diverse applications of DMDs in high-resolution imaging and microscopy, real-time holographic displays, data storage and retrieval, optical metrology and 3D measurements, and medical imaging and diagnostics.

15.1 High-resolution imaging and microscopy

DMDs have revolutionized high-resolution imaging and microscopy by providing dynamic, programmable light modulation. In DHM, DMDs are used to create complex illumination patterns that enhance the imaging process. DHM leverages the wavefront reconstruction capabilities of holography to provide detailed, three-dimensional images of microscopic samples. The high pixel density and fast switching speeds of DMDs enable rapid and precise control over the illumination, improving the resolution and quality of the captured images. This is particularly beneficial in biological research, where detailed visualization of cellular structures and dynamic processes is essential. Studies have demonstrated the effectiveness of DMDs in DHM and showcased the use of high-speed DMDs to achieve real-time holographic microscopy, enabling the capture of live cell dynamics with high temporal and spatial resolution. The programmability of DMDs also allows for advanced imaging techniques, such as structured illumination and phase shifting, further enhancing the capabilities of DHM systems [40].

15.2 Real-time holographic displays

Real-time holographic displays are one of the most promising applications of DMD technology. DMDs can generate dynamic holographic patterns at high speeds, enabling the creation of real-time 3D images and animations. This capability is critical for applications in entertainment, virtual reality (VR), and augmented reality (AR), where immersive and interactive 3D content is in high demand. The use of DMDs in holographic displays offers several advantages over traditional display technologies. The high switching speeds of DMDs allow for the rapid refresh rates required for smooth motion rendering, while the high resolution ensures detailed and lifelike images. Furthermore, the programmability of DMDs enables the generation of complex holographic scenes, enhancing the realism and depth of the displayed content. Research in this area has shown significant progress. A notable example is the development of holographic video displays that utilize DMDs to produce high-quality 3D video content. These systems level-rage the fast modulation capabilities of DMDs to update the holographic images in real time, creating a seamless viewing experience.

15.3 Data storage and retrieval

DMDs are essential in holographic data storage systems, providing high-density and high-speed data storage. These systems store data as holograms within a photo-sensitive material, allowing for precise control of light patterns. DMDs offer several benefits, including high resolution for recording large amounts of data in small volumes, fast switching speeds for quick access and retrieval, and programmability for generating multiple holograms in the same volume through techniques, such as angle and wavelength multiplexing. Advances in holographic data storage technology have demonstrated the potential of DMDs in this field and their efficiency in high-capacity data storage [41].

15.4 Optical metrology and 3D measurements

DMD-based holography systems are crucial for optical metrology and 3D measurements due to their precision and speed. They generate light patterns used in interferometry and holographic profilometry, which are essential for accurate surface and dimensional measurements. In interferometry, DMDs create reference wavefronts that interfere with reflected wavefronts, analyzing surface topography and thickness variations. In holographic profilometry, DMDs project structured light patterns onto the object's surface, capturing deformations captured by a camera to reconstruct the 3D shape with high precision. This technique is widely used in quality control, manufacturing, and materials science. Recent studies have demonstrated the effectiveness of DMDs in optical metrology, particularly in digital holographic interferometry, demonstrating their ability to perform high-precision measurements in real time [42].

15.5 Medical imaging and diagnostics

Digital 3D printing (DMDs) has revolutionized medical imaging and diagnostics by enabling high-speed and noninvasive imaging techniques. DMDs provide detailed visualizations of biological tissues and structures, making them crucial for applications, such as cellular imaging, tissue diagnostics, and surgical planning. In cellular imaging, DMDs capture detailed images of living cells without staining or labeling, providing real-time insights into biological processes and disease mechanisms. In tissue diagnostics, DMDs create high-resolution holographic images of tissue samples, enabling accurate diagnosis of abnormalities and diseases. The ability to capture detailed 3D information enhances diagnostic capabilities, enabling better assessment and treatment planning. DMD-based digital holographic microscopy has demonstrated its potential in biological research and medical diagnostics [43].

16. Conclusion

Digital holography, enhanced by the integration of DMDs, has shown significant advancements and promising potential across various applications. This study has explored the fundamentals of digital holography, the operating principles and advantages of DMDs, recent technological innovations, and a range of practical applications, including high-resolution imaging, real-time displays, data storage, optical metrology, and medical diagnostics. Detailed case studies and experiments highlight the practical effectiveness and versatility of DMD-based digital holography systems. The capability of high-speed and complex wavefront shaping using DMDs was demonstrated, with the DMD operating in a time-multiplexing mode to represent 8-bit grayscale images within 38.4 ms. This allows for precise focusing in live samples and other applications without compromising accuracy. For higher modulation rates, amplitude modulation was demonstrated, achieving a maximum frame rate for 8-bit images of 833 Hz, representing a 32-fold increase compared to time-multiplexing schemes. It can also be beneficial in applications where the projected pattern is time-integrated by light-sensitive materials, such as fluorescent molecules in imaging or photo-initiator molecules in 3D printing. These experiments demonstrated the successful reconstruction of holograms using DMDs and their applicability in holographic interferometric metrology. Future considerations include the potential use of gray-scale values through PWM of the micromirror flipping, ensuring enough temporal overlap of the micromirrors in the reflecting state to enable coherent superposition of the elementary spherical waves. Additionally, designing an optical arrangement to isolate and magnify a single replicated reconstruction is essential. A promising application is in digital holographic interferometry, where the numerical addition of stored and current digital holograms can be performed in near real-time, with the resulting sum hologram reconstructed by the DMD. This could enable real-time digital holography and allow DMD-reconstructed wavefields to be superimposed on optically generated ones for interferometric measurements. These advancements position DMD-based systems as a powerful tool for advancing the field of holography.

Acknowledgements

The author acknowledges the use of the EndNote tool for language polishing of the manuscript.

Author details

Muhammad Moin[1*], Mehrunisa Moin[1*], A. Qadoos[2], Abdul Waheed Anwar[2] and Junaid Hassan[2]

1 Institute of Fundamental and Frontier Sciences, University of Electronic Science and Technology of China, Chengdu, China

2 Department of Physics, University of Engineering and Technology, Lahore, Pakistan

*Address all correspondence to: moingcs786@gmail.com and mehrunisamoin3@gmail.com

References

[1] Suo J et al. Computational imaging and artificial intelligence: The next revolution of mobile vision. In: Proceedings of the IEEE. Vol. 111. IEEE; 2023. pp. 1607-1639

[2] Zhou J. Design and Simulation of Three-Dimensional Hologram Emitting Phased Arrays in the Near Field. Cambridge, MA: Massachusetts Institute of Technology; 2015

[3] Rabosh E et al. Quality control of the diffraction structure of display reflection holograms. Journal of Optical Technology. 2024;**90**(10):582-589

[4] Long O. Lensless Digital Holographic Microscopy for Differentiation and Counting of Blood Cells in a Microfluidic Device Using Computer Vision. York, UK: University of York; 2024

[5] Lai Y, Marquez M, Liang J. Tutorial on compressed ultrafast photography. Journal of Biomedical Optics. 2024;**29**(S1):S11524-S11524

[6] Khonina S, Kazanskiy N, Butt M. Exploring diffractive optical elements and their potential in free space optics and imaging-A comprehensive review. Laser & Photonics Reviews. 2024:2400377

[7] Conkey DB, Caravaca-Aguirre AM, Piestun R. High-speed scattering medium characterization with application to focusing light through turbid media. Optics Express. 2012;**20**(2):1733-1740

[8] Rhisheekesan A et al. Review on digital holography techniques using digital micromirror device. Optics and Lasers in Engineering. 2024;**177**:108120

[9] Lee B, Kim D, Lee S, Chen C, Lee B. High-contrast, speckle-free, true 3D holography via binary CGH optimization. Scientific Reports. 2022;**12**(1):2811

[10] Chen X, Zheng B, Liu H. Optical and digital microscopic imaging techniques and applications in pathology. Analytical Cellular Pathology. 2011;**34**(1-2):5-18

[11] Gabor D. Holography, 1948-1971. Science. 1972;**177**(4046):299-313

[12] Rosen J, Hai N, Bulbul A. Three-dimensional imaging using coded aperture correlation holography (COACH). Coded Optical Imaging. 2024;**6**:225-240

[13] Branch DEE. Optics & MEMS. Amsterdam, Netherlands: Elsevier; 1999

[14] Javidi B, Carnicer A, Anand A, Barbastathis G, Chen W, Ferraro P, et al. Roadmap on digital holography. Optics Express. 2021;**29**(22):35078-35118

[15] Zhuang Z, Ho HP. Application of digital micromirror devices (DMD) in biomedical instruments. Journal of Innovative Optical Health Sciences. 2020;**13**(06):2030011

[16] Lee S, Jo Y, Yoo D, Cho J, Lee D, Lee B. Tomographic near-eye displays. Nature Communications. 2019;**10**(1):2497

[17] Yoon T et al. Emerging applications of digital micromirror devices in biophotonic fields. Optics & Laser Technology. 2018;**104**:17-25

[18] Sandmeyer A et al. DMD-based super-resolution structured illumination microscopy visualizes live cell dynamics at high speed and low cost. BioRxiv. 2019:797670

[19] Escobar-Huertas J et al. Duchenne and Becker muscular dystrophy: Cellular mechanisms, image analysis, and computational models: A review. Cytoskeleton. 2024;**81**:269-286

[20] Kreis T, Aswendt P, Hö fling, R. Hologram reconstruction using a digital micromirror device. Optical Engineering. 2001;**40**(6):926-933

[21] Kumar M, Tounsi Y, Kaur K, Nassim A, Mandoza-Santoyo F, Matoba O. Speckle denoising techniques in imaging systems. Journal of Optics. 2020;**22**(6):063001

[22] Osten W, Kohler C, Liesener J. Evaluation and application of spatial light modulators for optical metrology. Óptica Pura y Aplicada. 2005;**38**(3):71-81

[23] Woo CM et al. Optimal efficiency of focusing diffused light through scattering media with iterative wavefront shaping. APL Photonics. 2022;**7**(4):1-11

[24] Luo Y et al. Focusing light through scattering media by reinforced hybrid algorithms. APL Photonics. 2020;**5**(1):1-12

[25] Curtis K et al. Holographic Data Storage: From Theory to Practical Systems. Hoboken, NJ: John Wiley & Sons; 2011

[26] Osten W et al. Recent advances in digital holography. Applied Optics. 2014;**53**(27):G44-G63

[27] Wu J, Feng T, Chen Q, Zuo C, Ma H. Photoacoustic guided wavefront shaping using digital micromirror devices. Optics & Laser Technology. 2024;**174**:110570

[28] Conkey DB et al. Genetic algorithm optimization for focusing through turbid media in noisy environments. Optics Express. 2012;**20**(5):4840-4849

[29] Zhao W, Wang Y, Zhang Z, Wang H. Multicriteria ship route planning method based on improved particle swarm optimization–genetic algorithm. Journal of Marine Science and Engineering. 2021;**9**(4):357

[30] Feng Q, Yang F, Xu X, Zhang B, Ding Y, Liu Q. Multi-objective optimization genetic algorithm for multi-point light focusing in wavefront shaping. Optics Express. 2019;**27**(25):36459-36473

[31] Hillman TR et al. Digital optical phase conjugation for delivering two-dimensional images through turbid media. Scientific Reports. 2013;**3**(1):1909

[32] Yu Z, Li H, Lai P. Wavefront shaping and its application to enhance photoacoustic imaging. Applied Sciences. 2017;**7**(12):1320

[33] Tamrin KF, Rahmatullah B. A review on noise suppression and aberration compensation in holographic particle image velocimetry. Cogent Physics. 2016;**3**(1):1142819

[34] Kreis T, Jüptner W, Schnars U. Telemetrology based on digital holography. In: Digital Holography and Three-Dimensional Display: Principles and Applications. 2006. pp. 213-234

[35] Lee J, Chae BG, Kim H, Yoon MS, Hugonnet H, Park YK. High-precision and low-noise dielectric tensor tomography using a micro-electromechanical system mirror. Optics Express. 2024;**32**(13):23171-23179

[36] Nikulin VV, Khandekar R, Bedi V. Application of holographic interferometry to in-plane and

out-of-plane misalignment and deformation measurements in packaging applications. In: Dimensional Optical Metrology and Inspection for Practical Applications IV. Bellingham, WA: SPIE; 2015. pp. 88-96

[37] Wang H, Wang X, Gao C, Wang Y, Zhao H, Yao Z. Multi-wavelength computational ghost imaging based on feature dimensionality reduction. In: Photonics. Vol. 11(8). Basel, Switzerland: MDPI; 2024. p. 739

[38] Khonina SN, Butt MA, Kazanskiy NL. A review on reconfigurable Metalenses revolutionizing flat optics. Advanced Optical Materials. 2024;**12**(14):2302794

[39] Khonina SN, Kazanskiy NL, Skidanov RV, Butt MA. Advancements and applications of diffractive optical elements in contemporary optics: A comprehensive overview. Advanced Materials Technologies. 2024:2401028

[40] Zehrer AC, Martin-Villalba A, Diederich B, Ewers H. An open-source, high-resolution, automated fluorescence microscope. eLife. 2024;**12**:RP89826

[41] Aman SS, Agbo DDA, N'guessan BG, Kone T. Design of a data storage and retrieval ontology for the efficient integration of information in artificial intelligence systems. International Journal of Information Technology. 2024;**16**(3):1743-1761

[42] Shaheen A, Bissacco G. Metrological characterisation of tool pre-setting optical systems based on 2D imaging. Measurement. 2024;**228**:114327

[43] Abhisheka B, Biswas SK, Purkayastha B, Das D, Escargueil A. Recent trend in medical imaging modalities and their applications in disease diagnosis: A review. Multimedia Tools and Applications. 2024;**83**(14):43035-43070

Chapter 5

Single-Pixel Digital Holography Using a Digital Micromirror Device

Ma. A. Araiza-Esquivel, Misael Torres-García, Diana I. Ortiz-Esquivel, Armando Santos-Amador, Raúl E. Balderas-Navarro, Alfonso López-Martínez and Humberto González-Hernández

Abstract

This chapter explores into the single-pixel (SP) digital holography (DH) technique, which employs a SP detector and a digital micromirror device (DMD) as a spatial light modulator (SLM). By projecting phase-encoded sampling patterns through computer-generated holograms (CGH), the system effectively measures and corrects wavefront aberrations, allowing for precise recovery of both amplitude and phase information from the sample. Building upon the advantages of single-pixel imaging (SPI), DH has shown interesting potential in applications such as optical scanning holography. Recent advancements have focused on improving reconstruction accuracy by combining complex-encoded masks and PS techniques. In addition, considering the technique as an SP camera offers benefits over traditional 2D-pixel array sensors, opening the door to exploring new digital holographic methods and applications. This chapter provides the essential steps for applying the SP holography technique and highlights its potential applications.

Keywords: spatial light modulation, single-pixel image technique, digital holography, Lee holograms, image processing

1. Introduction

Holography theory was first known in the late 1940s. Its inventor was working on electron microscopy issues when he developed the theory [1]. The importance of holography has been recognized for a long time, along with new technologies that have emerged, first with lasers and then with computers. However, to date, the potential of holography has yet to be fully applied. Early developments, such as SLMs, metasurface, monodetector, and faster cameras, paved the way for new applications of holography. With compressive sensing theory and neural networks (artificial intelligence), we have new prospects for exploiting its capabilities [2].

Figure 1(a) illustrates an analog holography interferometer. In this setup, the laser beam is split into two paths: one beam is directed toward the object, reflecting light

IntechOpen

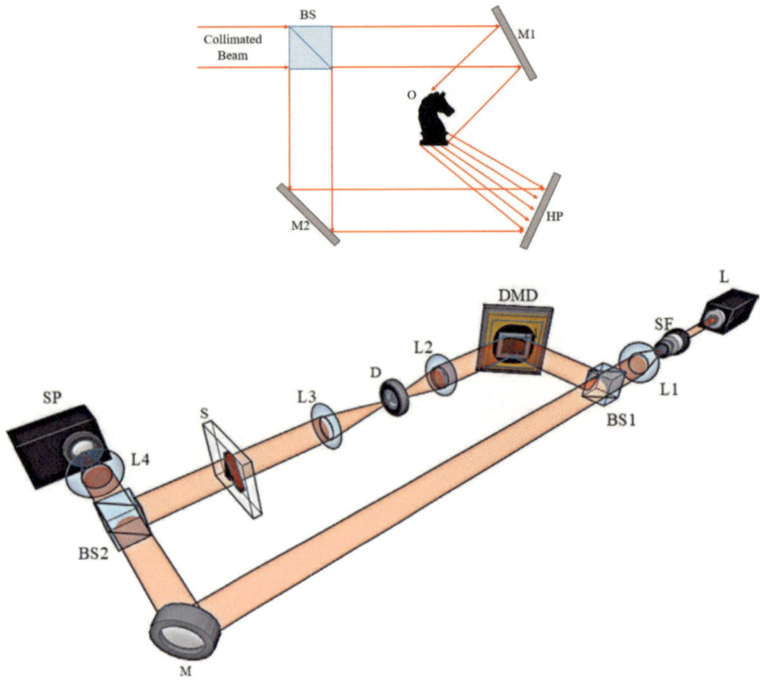

Figure 1.
Optical setups. (a) Analog holography: BS, beam splliter; M1, M2, mirrors; O, object; HP, holographic plate. (b) SP digital holography: L, laser; DMD, digital micromirror device; L1, L2, L3, L4, lenses; D, diaphragm; SF, spatial filter; S, sample; BS1, BS2, beam splitter; SP, single-pixel; M, mirror.

carrying all the information from the object's surface, while the other beam proceeds directly to the holographic plate to interfere with the object beam [3]. In traditional holography, the holographic plate functions similarly to photographic film but is composed of various chemical solutions tailored to the laser's wavelength used in recording. This technique has several limitations. However, with the introduction of digital cameras, the holographic plate has been effectively replaced, giving rise to digital holography.

DH recovers images of transparent or semitransparent objects without needing exogenous imaging contrast agents, even when these objects do not scatter light significantly. This technique retrieves the phase information with the interference between a wavefront scattered by an object and a reference wavefront. It typically uses a CCD sensor to record interference patterns. In the DH approach, a set of interference patterns are successively captured across the recording plane by shifting the phase of the object beam and keeping the reference beam unchanged [4]. Furthermore, DH has benefited from the advantages of SPI [5]. SPDH concerns the reconstruction of the amplitude and phase of an object by using complex-encoded masks together with the phase-shifting technique (PST) and exploiting the fact that the photodetector has no spatial resolution [6].

Advances in holography have introduced diverse methods for recording and reconstructing wavefronts. On the one hand, analog holography records interference patterns directly on a photosensitive material, preserving both phase and amplitude information. In contrast, DH uses the SP technique to capture data by modulating a laser beam through a DMD and measuring the corresponding intensities for each pattern with an SP detector (See **Figure 2**). The captured data are then processed through computational algorithms to reconstruct the hologram.

Analog holography offers high spatial resolution, making it ideal for high-fidelity imaging applications. However, it lacks flexibility in post-processing. SPDH, meanwhile, allows for adaptive modulation patterns but requires a high number of measurements to achieve comparable spatial resolution, which means a higher computational cost.

The SPI technique [7] recovers images of objects in situations where traditional cameras struggle to capture images, such as limited illumination or involving vision through scattering media [8]. This technique can work with light outside the visible spectrum [9]. Furthermore, SP imaging has been used in several applications, such as capturing object information in microscopy [10], hyperspectral imaging [11, 12], combined with compressive sensing techniques [7, 9, 13], and neural networks.

It is possible to use a balanced detector instead of a typical SP detector to obtain images of the object of interest, which can reduce the sampling and the processing time. This type of detection has been applied to incoherent Raman scattering technique [10], polarimetric measurements [14], optical coherence tomography [15], optical code-division multiple-access [16], computational imaging [17], and optical frequency domain imaging [18], among others.

SP technique uses SLMs to modulate the wavefront with sampling patterns and be recorded by the mono detector [2, 19]. Although the DMD is an amplitude modulator, it is possible to modulate the phase of wavefront employing holographic methods [20].

This chapter shows the fundamentals of SPDH employing a DMD as an SLM and a mono detector. We introduce the use of computer-generated Lee holograms and explore their application in subsequent optical processing using Hadamard basis functions.

The rest of this chapter is organized as follows. Section 2 provides an overview of SLM and DMD. Section 3 covers the theory behind generating Hadamard patterns. Section 4 explains the fundamentals of digital holography and the Lee method for generating holograms that encode the sampling patterns. Section 5 focuses on the single-pixel technique. Section 6 shows the optical setup and its descriptions. Section 7 shows the results. Section 8 concludes with key remarks and potential applications. Finally, Appendix A presents alternative sampling bases, such as cosine and noiselet functions, and Appendix B presents a flow chart with a general guide to applying the SP holographic method.

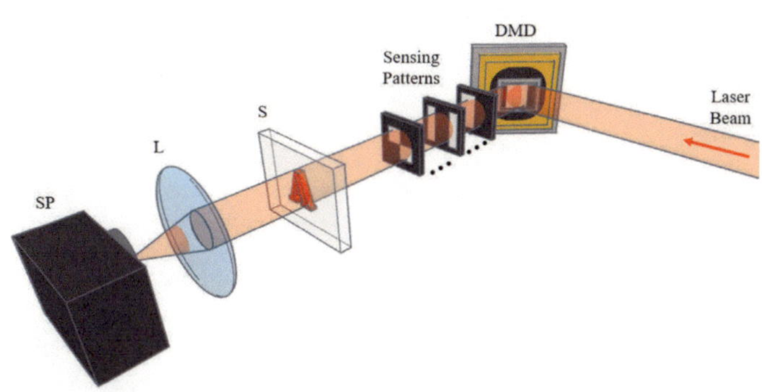

Figure 2.
SP technique: SP, single-pixel; L, lens; S, sample; DMD, digital micromirror device.

2. Spatial light modulators

SLM is a two-dimensional optical device, reflective or transmissive type, that modulates the properties of an incident laser beam (into the active surface of the device), such as phase, amplitude, and polarization, or it can even be a combination of any of these parameters. This capability of tailoring the light is known as structured light [21]. The SLM can be organized into two main categories based on their operational principles and applications, for example, (1) transmission or reflection, (2) modulate phase, amplitude, or polarization [22]. Different SLMs have been developed using diverse technologies, like liquid crystal SLMs, magneto-optic SLMs, deformable mirror SLMs, and acousto-optic Bragg cells. Usually, when referring to SLM, this mainly refers to those based on liquid crystals (LC). The LC molecules have a structure similar to an ellipsoid, and these molecules can be arranged in different configurations. The three phases of the LC are nematic, smectic, and cholesteric, being the nematic crystal the most used in SLMs [3].

2.1 DMD architecture

Among spatial modulators, the DMD stands out as a novel technology, initially developed for industrial applications by Texas Instruments in 1987 [23]. The DMD is a two-dimensional array of hundreds of thousands of micromirrors. Each micromirror operates independently, tilting between two states: $+12^\circ$ upward (state 1) and -12° downward (state 0) [24], as illustrated in **Figure 3**. A typical DMD features dimensions of approximately 1600×768 pixels, with micromirror pitches ranging from 7 to 14 micrometers. One of the key advantages of DMDs over other spatial light modulators (SLMs) is their ability to shape light beams at high speeds. They can display patterns at up to 27 kHz frequencies, enabling rapid beam manipulation. Another notable feature is the aluminum micromirrors, which provide high reflectance over a wide spectral range, spanning from 400 to 2000 nanometers [5, 25]. Due to its binary (1 or 0) programmable nature, the DMD is often referred to as a binary programmable mask, capable of reflecting and modulating light with predefined patterns.

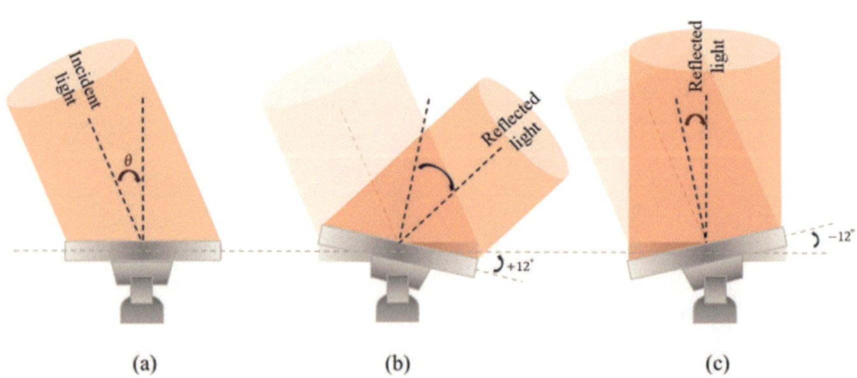

Figure 3.
DMD has two degrees of freedom $+12^\circ$, and -12°. (a) DMD is at rest, and the angle of the incident ray is $\theta = 24^\circ$. The reflected ray changes when the DMD is in (b) the on state and (c) the off state.

3. Sensing patterns generation: Hadamard basis

The Hadamard basis is used as a sample pattern in the SPDH theory. It is generated from the Hadamard-Walsh function. Due to their binary nature, Hadamard patterns can be easily implemented.

The Hadamard matrix of order N is a square matrix ($\mathcal{H}_N$), where entries take values of ± 1; this matrix is only defined for the dimension of $N = 2^k$. Two-dimensional forward Hadamard transform $\mathcal{H}\{\cdot\}$ is defined as follows:

$$\mathcal{H}\{f(x,y)\} = \sum_{x=0}^{N-1}\sum_{y=0}^{N-1} f(x,y)(-1)^{\sum_{i=0}^{k-1}(u_i\ x_i\ +\ v_i\ y_i)}, \tag{1}$$

where $f(x,y)$ represents a 2D image, (x,y) are spatial coordinates, and (u,v) are the coordinates in the Hadamard domain. Here $u_i,\ v_i,\ x_i$, and y_i are the binary representation of $u,\ v,\ x$, and y. The forward and inverse Hadamard transform satisfies the property of symmetric matrices and orthogonality.

Applying the inverse Hadamard transform to Eq. (3), the Hadamard sensing patterns are obtained:

$$P_H = \mathcal{H}^{-1}\{\delta(u,v)\}, \tag{2}$$

where δ function is defined as follows:

$$\delta(u,v) = \begin{cases} 1, & u = u_0, v = v_0 \\ 0, & \text{otherwise.} \end{cases} \tag{3}$$

Finally, each Hadamard pattern P_H is divided into two matrices: the positive pattern (P_{H^+}) and the negative pattern (P_{H^-}). **Figure 4** illustrates the application of the following equation:

$$P_H = P_{H^+} - P_{H^-}, \tag{4}$$

where

$$P_{H^+} = \frac{P_H + \mathbf{1}_N}{2}, \tag{5}$$

and

$$P_{H^-} = \frac{\mathbf{1}_N - P_H}{2}, \tag{6}$$

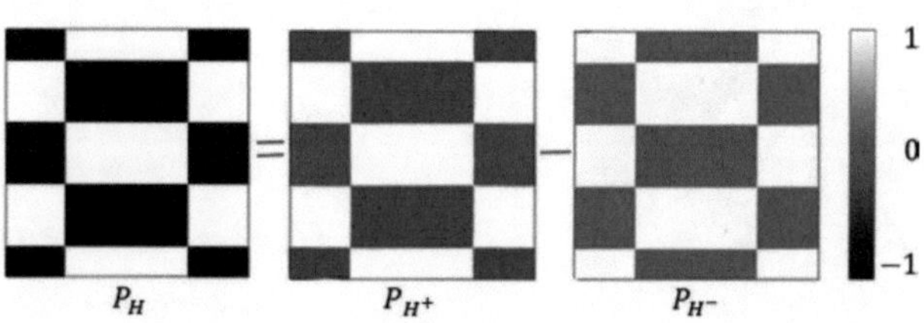

Figure 4.
Decomposition of values for P_{H^-} and P_{H^+} from P_H.

where $\mathbf{1}_N$ is a matrix of ones of order N [26]. In **Figure 5(a)**, the Hadamard patterns for $N = 3$ are shown.

The patterns are modified to ensure an accurate display so that the DMD can project them; each pattern is scaled eight times, from 64 x 64 to 512 x 512 pixels. Appendix A shows other sample patterns.

4. Holography and the Lee holograms method

Holography records the amplitude and phase of an object and reconstructs it into a 3D image. Such a process involves the interference between two coherent wavefronts: one coming from the object $(O(x,y) = A_O(x,y)e^{j(\phi_O(x,y)+\varphi)})$ and another from a tilted reference wavefront ($Re^{j\beta(x,y)}$), recording an interference pattern called hologram, that is given by [3]:

$$I_\varphi = |R|^2 + |A_O(x,y)|^2 + 2|R||A_O(x,y)|\cos[\beta(x,y) + \phi_O(x,y) + \varphi], \qquad (7)$$

where A_O and ϕ_O are the amplitude and phase of the diffracted light field by the object, respectively; R is the constant amplitude of reference beam; β is the carrier frequency; and $\varphi = 0, \pi/2, \pi, 3\pi/2$ is a constant phase. The four-phase PST consists of registering four intensities with a difference of $\pi/2$. We can derive the reconstruction of the object as follows:

$$O(x,y) = (I_0 - I_\pi) + j(I_{3\pi/2} - I_{\pi/2}), \qquad (8)$$

and the object phase as:

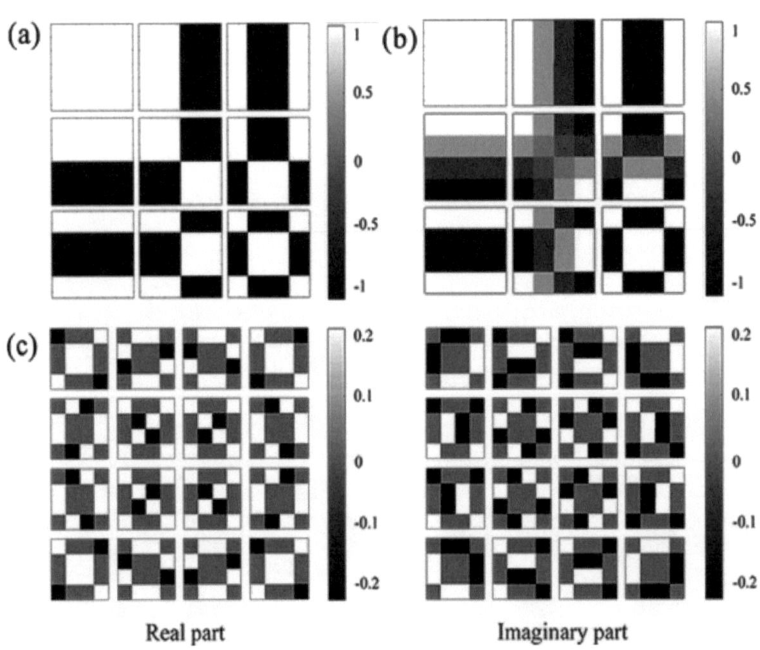

Figure 5.
(a) Hadamard and (b) cosine basis for $N = 3$, and (c) noiselets patterns for $N = 4$ (real and imaginary part).

$$\phi_O(x,y) = \tan^{-1}\frac{I_{3\pi/2} - I_{\pi/2}}{I_0 - I_\pi}. \tag{9}$$

4.1 Modulation pattern generation: Lee method

CGHs are numerically calculated and can encode objects without physical existence or complex mathematical functions. [27]. The reconstructed image can be recovered by displaying the hologram on the active surface of the modulator. Some of the CGH methods are the detour phase [28], the Lee holograms [29, 30], and the kinoforms [27], among others [31].

Applying the SP method to phase objects requires using complex-valued sampling patterns. These modulation patterns are generated using the Lee method, where Eq. (10) is derived by setting A_O and R equal to one in Eq. (7). The process for generating the modulating holographic patterns is given by:

$$I_\varphi \propto 1 + \cos[(x,y)\beta + \phi_O + \varphi]. \tag{10}$$

Here, we can consider adding the aberration correction as a phase contribution and replacing the phase ϕ_O with the sampling patterns P [32].

5. The single-pixel technique

In the SP imaging scheme, a set of patterns are projected onto an object, and an SP photodetector detects the light scattered by the object. Then, the electric signal from the detector is digitized by an analog-to-digital card converter connected to a computer. Let us consider the sampling patterns of microstructured light $P_i(x,y)$, with $i = 1,\ 2,\ \ldots, N$. The patterns are projected onto the object $O(x,y)$. The intensity distribution transmitted I_i through the sample for each sampling pattern $i - th$ and measured by the photodetector is given by:

$$I_i \propto \int\int |O(x,y)P_i(x,y)|^2 dxdy. \tag{11}$$

Subsequently, the object intensity is retrieved through a linear superposition of the recorded photocurrents with their respective sampling functions weighted as follows:

$$|O(x,y)|^2 = \sum_i^{N^2} I_i P_i(x,y). \tag{12}$$

The light intensity produced by the interference between a light beam coming from the object and a reference beam can be recorded by a photodiode. A set of four interferograms is generated by each pattern with a different phase shift of $\pi/2$ between them. To obtain a complete reconstruction, the object $O(x,y)$ is sampled with complex-values patterns $P_i(x,y)$, and a phase-shift φ is introduced in the object arm. The reference beam has a constant amplitude distribution R. Then, the irradiance measurements $I_{i,\varphi}$ corresponding to each interference pattern that is given by:

$$I_{i,\varphi} \propto \iint |O(x,y)P_i(x,y)\, exp\,(j\varphi) + R|^2 dxdy. \tag{13}$$

For each sampling pattern, four photocurrents $I_{i,0}, I_{i,\pi/2}, I_{i,\pi}$, and $I_{i,3\pi/2}$ are registered, one for each phase-shift (φ), aimed to apply the four-step PST. The measurements of the phase-shifted four photocurrents allow us to determine the complex coefficients associated with each sampling function projected onto the object. The complex coefficients are obtained according to the following expression, which adapts the conventional phase-shifting algorithm to SP digital holography:

$$c_i = \frac{1}{4|R|}\left[\left(I_{i,0} - I_{i,\pi}\right) + j\left(I_{i,3\pi/2} - I_{i,\pi/2}\right)\right], \tag{14}$$

where $j = \sqrt{-1}$. The complex amplitude distribution of the object is retrieved with the following expression:

$$O(x,y) = \sum_i^{N^2} c_i P_i(x,y). \tag{15}$$

Eq. (15) describes the object reconstruction as a linear combination of a basis function set weighted by their corresponding complex coefficients. The phase-encoded sampling functions used in our experiments are the Walsh-Hadamard functions. The Walsh-Hadamard functions have been implemented to sample an object with the SP DH. The sampling patterns are generated from the Hadamard-Walsh functions, representing modulation in horizontal and vertical directions, considering $+1$ or -1 values in the amplitude case and corresponding to 0 or π in the phase mode.

The method presented here is based on the approach outlined by Lee [30]. Phase information is codified into an amplitude binary hologram and displayed on a DMD screen. After illuminating the DMD screen, the light beam passes through a lens, and diffraction orders are produced. A diaphragm filters the $+1$st diffraction order at the focal distance. Again, the light passes through another second lens, and the phase information codified in the amplitude binary hologram is reconstructed at the focal distance. The optical system is capable of measuring the distortion inherent in the system. Using the DMD, the holographic patterns are projected onto the output of the 4f optical system, and the aberration is measured.

6. A typical SP-based optical setup

Figure 1(b) shows the experimental setup that consists of a Mach-Zehnder interferometer system. The light source is a He-Ne Gas laser (JDSU model 1508-1) with a wavelength of 633 nm. The laser passes through a spatial filter and a collimating lens. The laser is divided into a sample and a reference beam by a beam splitter. Then, the sample beam propagates to the DMD; the one used in this work is the DLP7000 model from Texas Instruments, with a resolution of 1024 × 768 pixels and a micromirror pitch of 13.7 μm. The DMD's active area reflects a structured wavefront based on the displayed mask. This modulated wavefront then travels through a 4f system consisting of two lenses: L2 (with a focal length of 250 mm) and L3 (with a focal length of 125 mm). The beam first propagates a distance f_2 to the Fourier plane,

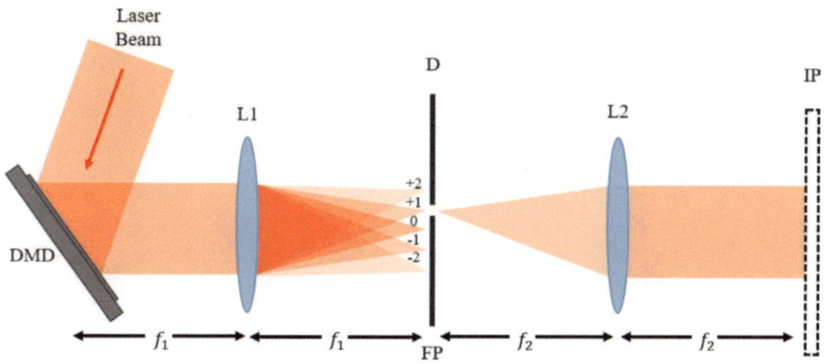

Figure 6.
Schematic setup for 4f system with DMD, digital micromirror device; L1, L2, lenses; D, Diaphragm; f1, f2, focal lengths; image plane; FP, Fourier plane.

where a diaphragm filters out unwanted frequencies. After filtering, the wavefront continues to reach lens L3. Upon passing through L3 and propagating an additional f_3, the beam reaches the image plane where a phase object was situated.

In **Figure 6**, a more detailed illustration of the 4f system is presented, highlighting its functionality in spatial frequency filtering. At the Fourier plane, a diaphragm is placed to selectively block higher-order diffraction frequencies generated by the structured wavefront from the DMD. By filtering out these undesired frequencies, the 4f system allows only the essential, lower-frequency components to pass through, ensuring a cleaner, more coherent wavefront for subsequent imaging and reconstruction. After passing through the phase object, the sample and reference beams merge at the beam splitter, producing an interference pattern. The Thorlabs $APD410A2/M$ photodetector then measures the intensity corresponding to each sensing pattern.

7. Results

Initially, a CCD camera is used to monitor the output scene. Four Lee holograms are generated to modulate the wavefront with a constant phase shift of $\pi/2$. The holographic patterns are displayed on the DMD screen, allowing the measurement of the aberration of the optical system without the object. **Figure 7(a)** shows the aberration of the system. After this measurement, the DMD projects four new holographic patterns with compensated aberration. **Figure 7(b)** shows the compensated aberration. Then, with these compensated holographic patterns, a sample is placed at the output plane of the 4f optical system. **Figure 8(a)** presents the object, while **Figure 8 (b)** and **(c)** show the retrieval amplitude and phase, respectively.

Afterward, the SP holographic method is applied: the CCD camera is replaced by a photodetector, and the Hadamard basis is encoded in the Lee holographic patterns for the sampling process. As described earlier, the first step is to measure the aberration of the optical system without the object. Then, the object is placed in the system and sampled using the new holographic patterns, which compensate in real time for the aberration of the system. **Figure 9** presents the retrieved amplitude and phase of an onion skin. Comparing the CCD camera with the SPI, we can see that the SPI can recover images with a resolution of 64×64 pixels while the CCD has a much higher resolution. On the other hand, notice that the amplitude image in **Figure 9(a)** is more pixelated than the recovered phase image in **Figure 9(b)**.

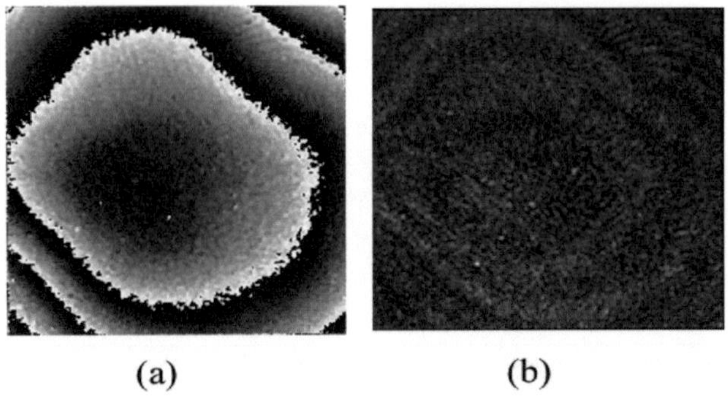

Figure 7.
Wavefront recorded with CCD camera and in the absence of an object: (a) aberration and (b) compensated aberration.

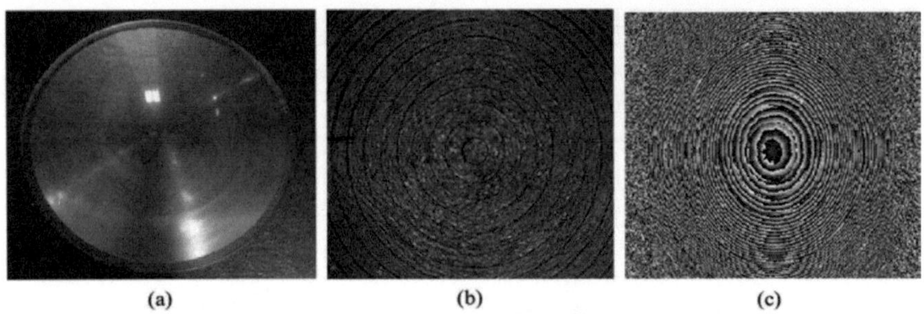

Figure 8.
Image recorded from a CCD camera. (a) Fresnel lens object, and reconstructed (b) amplitude and (c) phase image.

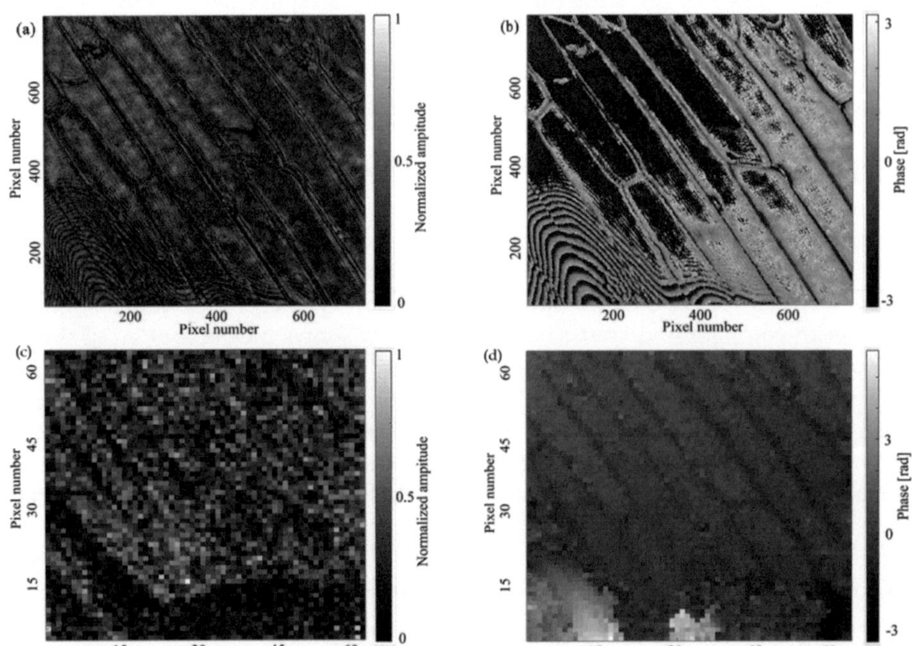

Figure 9.
Amplitude and Phase reconstructed image of an onion skin: (a) and (b) using a CCD camera, and (c) and (d) using the SP technique and Hadamard patterns.

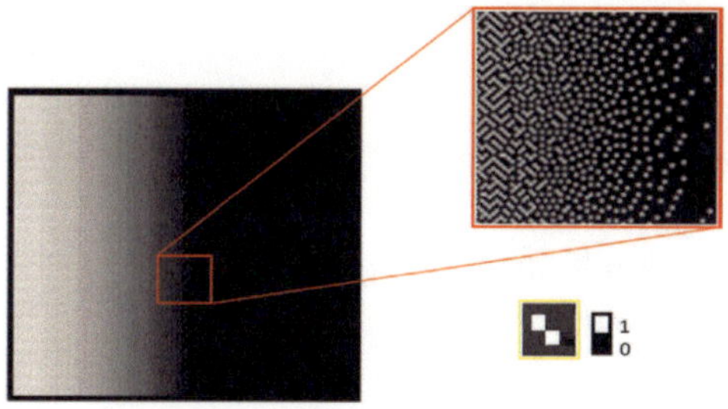

Figure 10.
Convert the image of the cosine pattern to a binary image.

Figure 11 in Appendix A shows a flow chart of the different stages of the technique of SP.

8. Conclusions

An SP holographic system using a DMD as a modulator and a mono detector has been presented. In summary, it concludes with the advantages and the limitations of the technique: (1) Although the DMD is an amplitude modulator, it was possible to phase modulate the incident wavefront. (2) The DMD is the fastest of the modulators. (3) The generated holographic patterns have codified three different phases. (4) It is not necessary to use a polarizer or a transductor to apply phase-shifting, avoiding a complex system. (5) The SP cameras offer benefits over traditional 2D-pixel array sensors, working in a broader range of wavelengths. (6) Due to the particularities of the SP imaging method, we infer that it is possible to obtain the phase of an object when a diffuse medium is placed between the object and the photodetector, so the presented optical system opens new perspectives that must be investigated. (7) The aberration correction can be made in real time. (8) Finally, the technique can be adapted for microscopy studies. (9) The technique requires longer data acquisition and signal processing time, but other technologies, methods, or theories can help to reduce time. The points discussed above are of great interest to other research areas.

Acknowledgements

One of the authors (M.T.G.) acknowledges partial financial support from CONAHCYT (Mexico).

Conflict of interest

The authors declare no competing interests.

Abbreviations

CGH	computer-generated hologram
DH	digital holography

DMD	digital micromirror device
LC	liquid crystal
PST	phase-shifting technique
SLM	spatial light modulator
SP	single-pixel
SPDH	single-pixel digital holography
SPI	single-pixel imaging

Nomenclature

$\otimes$	The Kronecker product
j	$\sqrt{-1}$
i	$1, 2, \ldots, N$
c	measured coefficients
δ	Delta function
φ	$0, \ \pi/2 \ , \pi, \ 3\pi/2$
π	3.1416
β	carrier frequency
θ	angle of incident wave
ϕ	phase of object wave
$f(x,y)$	2D image
A	amplitude of object wave
R	amplitude of reference wave
I	irradiance
P_{H^+}, P_{H^-}	positive and negative Hadamard patterns
P_{C^+}, P_{C^-}	positive and negative cosine patterns
$P^{Re}_{\mathcal{N}^+}, P^{Re}_{\mathcal{N}^-}$	positive and negative real part of noiselets patterns
$P^{Im}_{\mathcal{N}^+}, P^{Im}_{\mathcal{N}^-}$	positive and negative imaginary part noiselets patterns
(x,y)	spatial coordinates
(u,v)	frequency coordinates
$(l,\cdot)$	indicates the $l-th$ row vector of $\mathcal{N}_N$,
$(\cdot,t)$	refers to the t-th column vector of $N_{\mathcal{N}}$
$\mathcal{C}\{\cdot\}$	cosine transform
$\mathscr{H}\{\cdot\}$	Hadamard transform
$\mathcal{N}\{\cdot\}$	noiselets transform

Appendix A. Generation pattern: Cosine and noiselets

A.1 The cosine transform

Two-dimensional cosine transform $\mathcal{C}\{\cdot\}$ for an image $f(x,y)$ of size $N \times N$ is defined as:

$$F(u,v) = \sum_{x=0}^{N-1}\sum_{y=0}^{N-1} \alpha(u)\alpha(v)f(x,y)\cos\left[\frac{\pi u\ (2x+1)}{2N}\right]\cos\left[\frac{\pi v\ (2y+1)}{2N}\right], \tag{16}$$

$$f(x,y) = \sum_{x=0}^{N-1}\sum_{y=0}^{N-1} \alpha(u)\alpha(v)F(u,v)\cos\left[\frac{\pi u\ (2x+1)}{2N}\right]\cos\left[\frac{\pi v\ (2y+1)}{2N}\right], \tag{17}$$

where $\alpha(u=0) = \sqrt{\frac{1}{N}}$ and $\alpha(u \neq 0) = \sqrt{\frac{2}{N}}$; $\alpha(v) = \sqrt{\frac{2}{N}}$ for $n = 1,2,3, \ldots, N-1$, and $\alpha(0) = \sqrt{\frac{1}{N}}$.

The cosine sensing patterns can be generated by applying an inverse cosine transform to a delta function defined by Eq. (3):

$$P_C = C^{-1}\{\delta(u,v)\}. \tag{18}$$

Analogous to the process implemented in the Hadamard transform $P_C(x,y)$ is divided into a positive and negative matrix. Therefore, it can be expressed as:

$$P_C = P_{C^+} - P_{C^-}. \tag{19}$$

These patterns display a gradient, thus requiring a binarization process to ensure their correct representation on the DMD. A Floyd-Steinberg error diffusion algorithmis applied for this process, resulting in a dithered image. See **Figure 10**. **Figure 5(b)** shows the patterns for the cosine basis $N = 3$.

A.2 The forward noiselets transform

Generating the noiselets matrix requires an iterative process, starting with a 1×1 matrix $\mathcal{N}_1 = [1]$ and continuing to write N-order noiselets matrix using the following equations:

$$\mathcal{N}_N(l,\cdot) = \frac{1}{2}(1-j, 1+j) \otimes \mathcal{N}_{N/2}\left(\frac{l}{2},\cdot\right), \quad l = 0,2,4, \ldots, N-2 \tag{20}$$

$$\mathcal{N}_N(l,\cdot) = \frac{1}{2}(1+j, 1-j) \otimes \mathcal{N}_{N/2}\left(\frac{l-1}{2},\cdot\right), \quad l = 1,3,5, \ldots, N-1, \tag{21}$$

where $(l,\cdot)$ indicates the $l-th$ row vector of $\mathcal{N}_N, j = \sqrt{-1}$ and $\otimes$ is the Kronecker product. The pattern $P_{\mathcal{N}}(x,y)$ is generated appling the noiselets transforms to Eq. (3). In this case, the noiselets matrix contains complex numbers, as shown in the example in **Figure 5(c)**. To display them correctly by DMD (in binary elements), the patterns are normalized $[1, -1, j(1), -j(1)]$. Thus, the complex function is decomposed into real, imaginary, positive, and negative components, as follows:

$$P_{\mathcal{N}} = P^{Re}_{\mathcal{N}^+} - P^{Re}_{\mathcal{N}^-} + j\ P^{Im}_{\mathcal{N}^+} - j\ P^{Im}_{\mathcal{N}^-}. \tag{22}$$

where the super index indicates the positive and negative part in the noiselets matrix.

Appendix B. Process flowchart

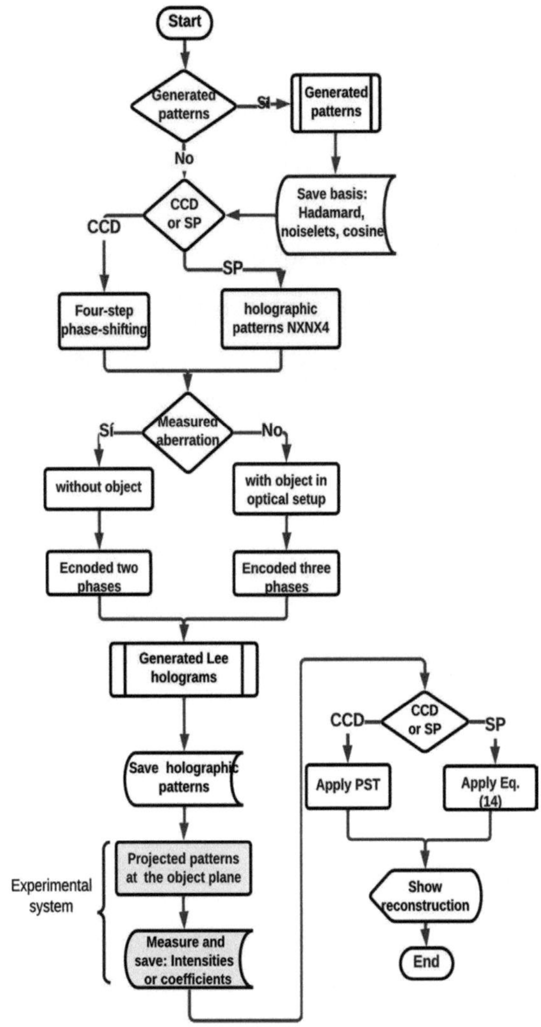

Figure 11.
Flow chart of the SP holographic technique.

Author details

Ma. A. Araiza-Esquivel[1*†], Misael Torres-García[2†], Diana I. Ortiz-Esquivel[1†], Armando Santos-Amador[2†], Raúl E. Balderas-Navarro[2†], Alfonso López-Martínez[1†] and Humberto González-Hernández[1†]

1 Unidad Académica de Ingeniería Eléctrica, Benemérita Universidad Autónoma de Zacatecas, Jardín Juárez 147, Zona Centro, Zacatecas, México

2 Instituto de Investigación en Comunicación Óptica (IICO-UASLP), San Luis Potosí, México

*Address all correspondence to: araizama@uaz.edu.mx

† These authors contributed equally.

References

[1] Gabor D. A new microscopic principle. Nature. 1948;**161**:777-778

[2] Javidi B, Carnicer A, Anand A, Barbastathis G, Chen W, Ferraro P, et al. Roadmap on digital holography. Optics Express. 2021;**29**:35078

[3] Goodman JW. Introduction to Fourier Optics. 3rd ed. US: Roberts and Company Publishers; 2005

[4] Yamaguchi I. Phase-Shifting Digital Holography: Principles and applications. In: Poon TC, editor. Digital holography and three-dimensional display: principles and applications. Boston, MA: Springer US; 2006. pp. 145-171. DOI: 10.1007/0-387-31397-4_5

[5] Gibson GM, Johnson SD, Padgett MJ. Single-pixel imaging 12 years on: A review. Optics Express. 2020;**28**(19): 28190-28208

[6] Martínez-León L, Clemente P, Mori Y, Climent V, Lancis J, Tajahuerce E. Single-pixel digital holography with phase-encoded illumination. Optics Express. 2017;**25**(5):4975-4984

[7] Duarte MF, Davenport MA, Takhar D, Laska JN, Sun T, Kelly KF, et al. Single-pixel imaging via compressive sampling. IEEE Signal Processing Magazine. 2008;**25**:83-91

[8] Tajahuerce E, Durán V, Clemente P, Irles E, Soldevila F, Andrés P, et al. Image transmission through dynamic scattering media by single-pixel photodetection. Optics Express. 2014;**22**:16945

[9] Watts CM, Shrekenhamer D, Montoya J, Lipworth G, Hunt J, Sleasman T, et al. Terahertz compressive imaging with metamaterial spatial light modulators. Nature Photonics. 2014;**8**: 605-609

[10] Molotokaite E, Kumar V, Manzoni C, Polli D, Cerullo G, Marangoni M. Raman-induced Kerr effect microscopy with balanced detection. Journal of Raman Spectroscopy. 2013;**44**(10):1385-1392

[11] Suo J, Wang Y, Bian L, Dai Q. Single-pixel hyperspectral imaging. In: Proceedings of the SPIE Optoelectronic Imaging and Multimedia Technology IV. Vol. 10020. Beijing, China: SPIE; 2016. p. 38-44

[12] Bian L, Suo J, Situ G, Li Z, Fan J, Chen F, Dai Q. Multispectral imaging using a single bucket detector. Scientific Reports. 2016;**6**:24752

[13] Mardani D, Kondakci HE, Martin L, Abouraddy AF, Atia GK. Compressive optical interferometry under structural constraints. Optics Express. 2018;**26**(5): 5225-5239. Available from: https://opg.optica.org/oe/abstract.cfm?URI=oe-26-5-5225

[14] Wang X, Cui Y, Sun W, Ye J, Zhang Y. Terahertz polarization real-time imaging based on balanced electro-optic detection. Journal of the Optical Society of America A. 2010;**27**:2387

[15] Bradu A, Podoleanu AG. Fourier domain optical coherence tomography system with balance detection. Optics Express. 2012;**20**:17522

[16] Tseng SP. Modified multiphotodiode balanced detection technique for improving SAC-OCDMA networks. Optics Communications. 2015;**344**:38-42

[17] Soldevila F, Clemente P, Tajahuerce E, Uribe-Patarroyo N, Andrés P, Lancis J. Single-pixel imaging using balanced detection and a digital micromirror device. In: Proceedings of

the SPIE Emerging Digital Micromirror Device Based Systems and Applications X, San Francisco, California, US. Vol 10546. SPIE; 2018. p.1054604. DOI: 10.1117/12.2289821

[18] Chen Y, de Bruin DM, Kerbage C, de Boer JF. Spectrally balanced detection for optical frequency domain imaging. Optics Express. 2007;**15**:16390

[19] Endo Y, Nakajima G. Compressive phase object classification using single-pixel digital holography. Optics Express. 2022;**30**(15):28057-28066. Available from: https://opg.optica.org/oe/abstract.cfm?URI=oe-30-15-28057

[20] Jiao S, Zhang D, Zhang C, Gao Y, Lei T, Yuan X. Complex-amplitude holographic projection with a digital micromirror device (DMD) and error diffusion algorithm. IEEE Journal of Selected Topics in Quantum Electronics. 2020;**26**(5):1-8

[21] Forbes A. Structured light, Tailored for Purpose. Optics and Photonics New. 2020;**31**(6):24-31

[22] Jullien A. Spatial Light Modulators. Photoniques. 2020;**1**(101):59-64

[23] Sampsell JB. Digital micromirror device and its application to projection displays. Journal of Vacuum Science & Technology B: Microelectronics and Nanometer Structures Processing, Measurement, and Phenomena. 1994; **12**(6):3242-3246

[24] GmbH VMB. SuperSpeed Specification [Internet]. (n.d.) Available from: https://www.vialux.de/en/superspeed- specification.html [Accessed: November 4, 2024]

[25] Quijada MA, Travinsky A, Vorobiev D, Ninkov Z, Raisanen A, Robberto M, et al. Optical evaluation of digital micromirror devices (DMDs) with UV-grade fused silica, sapphire, and magnesium fluoride windows and long-term reflectance of bare devices. In: Proceedings of the SPIE Advances in Optical and Mechanical Technologies for Telescopes and Instrumentation II. Edinburg, UK: SPIE; Vol. 9912. 2016. p. 1756-1765. DOI: 10.1117/12.2233341

[26] López-García L, Cruz-Santos W, García-Arellano A, Filio-Aguilar P, Cisneros-Martínez JA, Ramos-García R. Efficient ordering of the Hadamard basis for single pixel imaging. Optics Express. 2022;**30**(8):13714-13732

[27] Tricoles G. Computer generated holograms: An historical review. Applied Optics. 1987;**10**(26):4351

[28] Lohmann AW, Paris D. Binary Fraunhofer holograms, generated by computer. Applied Optics. 1967;**6**(10): 1739-1748

[29] Lee WH. Binary synthetic holograms. Applied Optics. 1974;**13**(7):1677

[30] Lee WH. III Computer-Generated Holograms: Techniques and Applications. In: Progress in Optics. Vol. 16. Elsevier; 1978. p. 119-232. Available from: https://www.sciencedirect.com/science/article/pii/S0079663808700726

[31] Georges T, Nemetallah RA, Williams LA. Analog and Digital Holography with MATLAB. USA: Society of Photo-Optical Instrumentation Engineers (SPIE); 2015. p. 528. Available from: https://la.mathworks.com/academia/books/analog-and-digital-holography-with-matlab-nehmetallah.html

[32] González H, Martínez-León L, Soldevila F, Araiza-Esquivel M, Lancis J, Tajahuerce E. High sampling rate single-pixel digital holography system employing a DMD and phase-encoded patterns. Optics Express. 2018;**26**(16): 20342-20350

Chapter 6

Controlled Phase Shifting in Reflective Digital Holographic Microscopy

Diego Palacios and Michael R. Wang

Abstract

Digital holographic microscopy (DHM) with 3D microscopic imaging capability is attractive for biological sampling, particle identification, and surface topographic profiling. The Michelson (Linnik) optical configuration along with microscope objectives is used to create a hologram with the reflections of a sample of 3D spatial structures. Instead of instantaneous volume recording of the hologram in a holographic film, a digital camera is used to sequentially record four phase-step holograms in 2D format, namely the digital holograms (DHs). A piezoelectric actuator was added to implement the formation of the four-phase-step DHs. Using a four-image phase-shifted algorithm, the 3D microscopic image of the object can be extracted. The recording of the DHs is fast, 28 ms, and the phase retrieval computation is fast as well, 24 ms. The DHM has been utilized to examine the surface topography of a lithography sample showing field of view of 312 μm × 234 μm, axial resolution of 3.31 nm, lateral resolution of 488 nm, and a maximum topography height of 270 nm.

Keywords: digital holographic microscopy, interferometric microscopy, holograms, phase shifting, surface topography

1. Introduction

Holograms are dependent on the coherent properties of light and the interference patterns that are created when two coherent lights interact with one another. In traditional analog holography (TAH), the intensity and phase information of the interference pattern is stored in a volume light-sensitive holographic medium, which can be a crystal or a film. A reference light source similar to that used in the hologram recording can reveal, or readout, the stored 3D information of the object for observation and display. TAH is excellent in its instantaneous recording of the 3D information of the object. However, it suffers from the limited available crystal or film recording media, hologram fixing post-processing, hologram readout erasure degradation (volatility), and hologram storage lifetime limitation. The hologram recording may also be extremely vibration-sensitive while the post-processing may be time consuming. Recorded holograms are not possible for online delivery to remote interested users or observers.

The DH is also recorded using coherent lights onto a 2D digital camera as the analog recording medium. While the interference amplitude and phase information are recorded in the 2D camera pixels, the recording loses the angular interference fringe information in the direction perpendicular to the plane of the camera array pixels. Successively recording a few DHs with the tuning of the position of the camera pixels or the tuning of the recording reference light phase can help regain the lost information for the hologram readout through a phase retrieval computation algorithm. Since the camera pixels are highly sensitive to light, the recording of each DH and a set of DHs, even with reference light tuning, is fast, 7 ms per hologram in our experiments. With the elimination of hologram fixing post-processing in TAH, the DHs are stable (not volatile to readout degradation) and are friendly to online electronic delivery. The phase retrieval computation is fast, 24 ms, for recovery of the 3D information of the object. Although this computation readout is not as fast as the light illumination readout in a TAH, the recovered 3D image with user-friendly rotation and display enhances its practical significance. The use of DH technology in a microscope configuration would thus make it possible for the implementation of DHM to facilitate the examination of biological and particle samples and surface topographic profiling.

Due to the formation of the hologram by interference between a reference beam and an object beam from a common coherence light source [1], the acquisition of a set of digital holograms for microscopy through the tuning of the reference beam phase makes the DHM similar in terminology and optical setup as optical interferometric microscopy (OIM). The DHM can be configured to operate in transmission with a Mach Zehnder interferometer to image specimens such as dielectrics [2] or biological samples including cell cultures [3–5]. Reflective DHM is convenient for imaging objects that are limited by the numerical aperture. For instance, topographic and tomographic imaging of live species such as tadpole eyes [6], several layers of onion cells [7], or *in vivo* human skin [8] has been shown with a reflective DHM. Research has been focused on improving measurement sensitivity [9], exploration of surface topography applications [9], efficient algorithm for extraction of interferometric phase information [10], and exploration of full-field optical coherence tomography [11].

In this book chapter, we describe the concept of DHM in a reflective configuration. The Michelson interferometer optical system with a Linnik configuration, the coherence light source requirement, interferometric phase tuning, DH acquisition, phase extraction algorithm, measurement sensitivity and resolution, and a practical application of the DHM for lithographic 3D surface profiling are presented.

1.1 Optical microscope and Linnik interferometric microscope configuration

A conventional optical microscope consists of a white light source (typically a tungsten halogen lamp) with a collimation lens, optical beam splitters, objective lens, ocular lens (or eyepiece), optical camera, and a sample holding platform as illustrated in **Figure 1(a)**. The collimated white light is partially directed downward by the beam splitter to the objective lens that focuses the beam and illuminates the sample. The other portion of the split light is unused and is blocked. The reflected beam from the sample is then imaged to the position near the ocular lens front focal plane, and the image is enlarged by the ocular lens to the observation eye. This image beam can also be directed to an optical camera typically mounted on the top to provide a direct view of the microscope image of the sample on a display screen connected to the camera. The sample holding platform provides needed translation and focusing adjustment of the sample position.

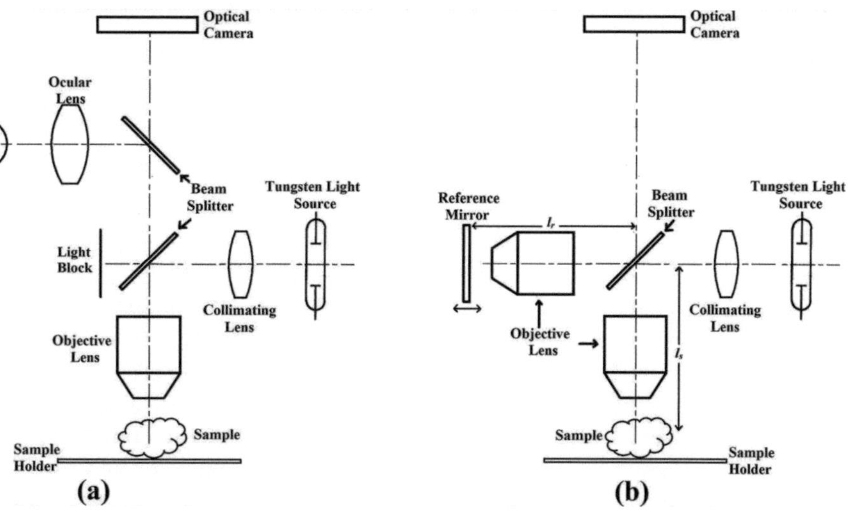

Figure 1.
Schematic of a conventional optical microscope (a) and a Linnik interferometric microscope configuration (b).

Let the illumination power of the light beam before the beam splitter be I_0 and the beam splitter ratio be 50:50. Ignore the power loss by the lens and let the local (corresponding to one image pixel on the camera) sample reflection coefficient be r_s. The optical power I_s received by the camera pixel is thus

$$I_s = 0.25 I_0 r_s \tag{1}$$

By converting captured light power for all image pixels for image display, we can obtain the microscope image of the sample.

Clearly, the conventional microscope image is only dependent on the light reflection coefficient r_s of the sample surface. Within the depth of focus of the objective lens, the microscope image is a 2D image of the sample surface. The topographic information of the sample surface is lost. The microscope lateral image resolution is closely related to the magnification power of the objective lens (assuming no other imaging lens is involved) and the pixel dimension of the camera. The field of view (FOV) of the microscope is also a function of the camera pixel array dimension if the optical imaging does not have an iris power cutoff to the imaging beam. The unused split beam power is wasted.

As illustrated in **Figure 1(b)**, if we utilize the previous unused split beam in a near identical optical configuration of the sample beam with a mirror of reflection coefficient r_m, and assume the mirror reflected power is overlapped with that from the sample local spot at the same pixel of the camera, the optical power I_s received by the camera pixel is now

$$I_s = 0.25 I_0 \left[(r_s + r_m) + \sqrt{r_s r_m} e^{-(l_r - l_s)^2 \Delta k^2} cos\left[2k_0 (l_r - l_s) \right] \right] \tag{2}$$

The mirror-reflected beam is considered as a reference beam. The addition of the sample beam (from the sample reflection) and this reference beam is not a simple numerical power addition, but it is an optical path length difference δ dependent interference. The difference between the sample beam path length l_s and the reference beam path length l_r is δ. Note that we use beam path length (instead of physical length) to include the effect of refractive indices of optical elements. Because of the optical path length difference δ, the combined beam can have constructive interference with a higher power or destructive interference with a lower power at the camera sensor pixel. This optical path length difference is also a phase factor in Eq. (2). Obviously, by evaluating this phase factor of the acquired image, we can obtain the sample path length l_s from a known reference beam path length l_r. Since l_s is sample surface height dependent, the pixel dependent l_s mapping would reveal the 3D topographic information of the sample so long as there is prior knowledge of the light source center wave number k_0 and spectral bandwidth in wavenumber Δk. This would make the optical system as illustrated in **Figure 1(b)** a practical 3D microscope.

The two-beam optical interference configuration, as illustrated in **Figure 1(b)**, is the Linnik configuration. Through the beam splitter, the combined sample beam and reference beam at the camera could be on-axis parallel or with a small angle if we slightly tilt the reference mirror. A tilted angle configuration is used in conventional two-beam transmission hologram recording since the interference fringes can be volume recorded in the holographic film. Using a camera with 2D sensing pixels for our recording of pixel-dependent $I_s(x, y)$, we are only able to record the interference fringe (digital hologram) in a given plane (x–y plane) while the volumetric fringes in the direction z perpendicular to the camera sensor plane is lost. In principle, we could shift the camera sensor plane along this z axis to acquire a few image $I_s(x, y)$ to obtain the 3D information $I_s(x, y, z)$ which could then be used to extract the 3D surface information or surface height information $h(x, y)$ which is a subtraction of $l_s(x, y)$ to a common average l_s.

While the repositioning of the camera for acquiring a set DH $I_s(x, y)$ at different z is possible, it would be more convenient to reposition the reference mirror, which is lighter than the camera, altering the reference distance l_r. A piezoelectric actuator (PZA) is adopted for the modulation of l_r.

In principle, the smaller the camera sensing pixel dimension, the better the lateral resolution of the 3D image $h(x, y)$. The more shifted positions of l_r, the better the height h resolution. In this book chapter, we will show that four positions of l_r would facilitate sufficient height resolution of the 3D microscope. With only four reference mirror positions the acquisition of a set of 4 DHs will be fast, making the system less sensitive to environmental changes such as vibration. The phase extraction computation recovery of the 3D image $h(x, y)$ will be fast as well.

Because of the two-beam interference, the optical path length difference δ should be within the coherence length of the light source. We thus must examine the conventional tungsten halogen light source on its coherence length and show its validity in this Linnik DHM configuration.

1.2 Coherence length of the light source

Based on the schematic design of DHM as illustrated in **Figure 1(b)**, we constructed a compact DHM as shown in the 3D drawing in **Figure 2**. Here, the fiber-coupled tungsten halogen light source (not shown) is used and the illumination beam is collimated by a fiber collimator lens at right. The beam splitter sends the split beams to a measurement objective lens and sample at bottom and to another reference objective lens and reference mirror at left. The reference mirror is mounted on a PZA for minor positional shifting modulation. The sample and reference beams are combined at top and are imaged to a CMOS camera for recording of the digital hologram. Each recorded DH is associated to one position l_r of the reference mirror. Four DHs at

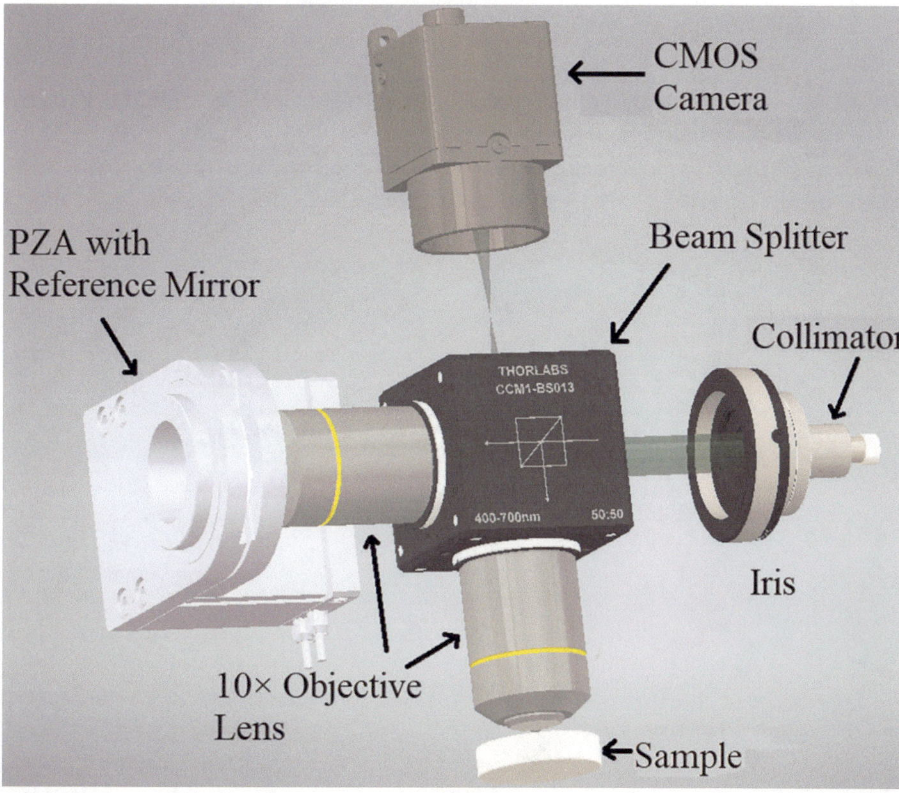

Figure 2.
3D schematic of a constructed digital hologram microscope in Linnik configuration. The constructed system dimensions are 10″ in height, 10″ in width, and 3.5″ in depth.

four I_r will be recorded as a set for the phase extraction computation of the 3D topographic surface information $h(x, y)$ of the sample.

A laser with a long coherence length could be used as a light source. To minimize speckle formation and minimize alteration to the conventional microscope system, we consider the traditional tungsten halogen lamp as the light source. To improve the light beam illumination uniformity, a fiber-coupled tungsten halogen lamp is adopted. To improve the coherence length of the light source, a spectral filter at center wavelength λ_0 of 540 nm, and spectral bandwidth $\Delta\lambda$ of 40 nm is inserted at the lamp housing before the multimode fiber (400 μm core diameter) coupling to the collimator. For lateral sample surface imaging with a fine resolution, a 10× microscope objective is used to magnify the sample surface to the CMOS camera, performing the function similar to a conventional microscope. The lateral resolution is thus determined by the optical magnification and the pixel density of the camera. To match with the sample arm, an identical 10× microscope objective is added in the reference arm to minimize system chromatic aberration [12].

With the use of the fiber-coupled light source and the spectral filter, the interference fringes acquired by the CMOS camera are shown in **Figure 3(a)** and **(b)** for a sample partially coated with a 125 nm thick chromium on fused silica. The darker and brighter portions are the fused silica and chromium-coated surfaces caused by the lower and higher reflectivity, respectively. **Figure 3(c)** shows the interference as a function of the optical path length difference δ. When the optical path length

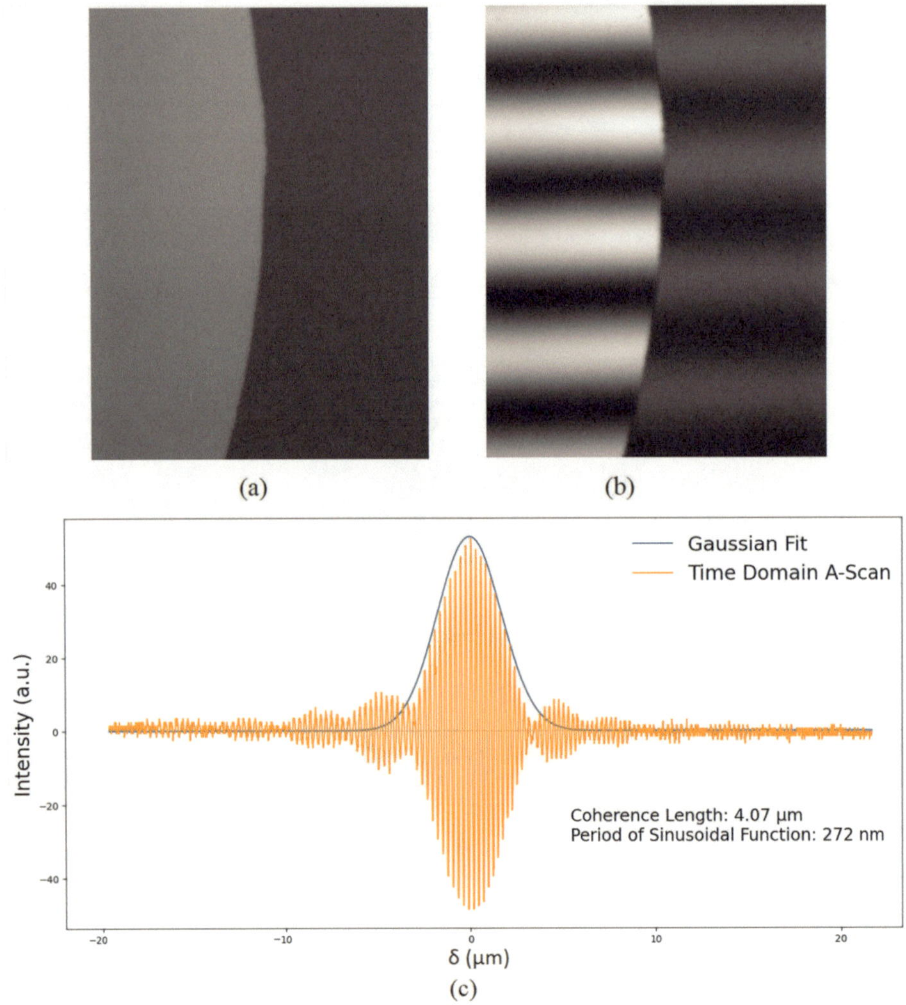

Figure 3.
Image of the sample when there is no interference (a) and with interference (b). Interference evaluation of the coherence length shows strong interference when the optical path length difference δ is near zero (c).

difference is near zero, there is strong interference, as observed in **Figure 3(c)**. The coherence length is determined as the width of this strong interference, which is about 4.07 µm. When the optical path length difference is larger than the coherence length, the interference is diminished as observed in **Figure 3(a)**. This experiment confirmed that the traditional tungsten halogen lamp with a multimode fiber coupler and a spectral filter of 40 nm bandwidth at 540 nm center wavelength is a practical and useful light source for this DHM configuration.

The coherence length l_c of a light source is typically evaluated by equation

$$l_c = \frac{2 \ln 2}{n\pi} \frac{\lambda_0^2}{\Delta\lambda} \tag{3}$$

where n is the refractive index of the medium. Eq. (3) is valid for the Gaussian spectral intensity distribution. Using the spectral filter with 540 nm center wavelength and

40 nm spectral bandwidth at n of 1, the coherence length l_c is calculated to be 3.22 μm. The experimentally evaluated coherence length is 4.07 μm because the spectral intensity distribution is not Gaussian [13]. Thus, when selecting a filter of a different center wavelength and spectral bandwidth, the calculated coherence length may not be exact, but it would provide a reasonably good estimation of the coherence length.

So long as the optical path length difference between the reference mirror and measurement sample is within the measured l_c effective DHM measurements can be made. With high-quality DHs such as that presented in **Figure 3(b)**, it is possible to modulate the reference mirror positions and thus the interference pattern phase and create a direct relationship between DH $I_s(x,y)$ and reference mirror position l_r [11]. The result is the creation of a phase map whose maximum unwrappable height h range is $\lambda_0/2$. Heights that exceed this value, while they still create a DH, will have added 2π phase that cannot be uniquely phase unwrapped [14].

Figure 4 contains DHs E_1, E_2, E_3, and E_4 of a surface step acquired at relative phases φ of 0, π/4, π/2, and 3π/4, respectively, where $\varphi = 2k_0\delta$ and $k_0 = \frac{2\pi}{\lambda_0}$. It shows that the interval between images is π/4 through the vertical motion of the interference fringes with a horizontal orientation. The acquisition of the 4 DHs has been performed by having the reference mirror slightly tilted to introduce the interference fringes needed for detection of phase change through the motion of the fringes. Fringe orientation is found to have minimal effect on phase mapping and subsequent phase unwrapping [15].

Each DH follows the time-domain low-coherence interference equation $I_s(x,y) = \bar{I}(x,y) + A(x,y)\cos(\varphi(x,y))$ [11, 12]. The coefficient $\bar{I}(x,y)$ is the constant average intensity delivered to the CMOS camera pixels while $A(x,y)$ is the coherent envelope. The phase map and height information are related to each other according to $\varphi(x,y) = 2k_0\delta(x,y)$ [12]. Variables x and y represent pixel locations on the camera. The relationship between phase and height is rearranged to provide the following:

$$\delta(x,y) = \frac{\varphi(x,y)}{2k_0} = \frac{\varphi(x,y)\lambda_0}{4\pi} \tag{4}$$

With the relationship between phase, wavelength, and optical path length difference, it can be determined that for a π/4 phase change a 33.75 nm optical path length change must be induced. The phase-shifted DHs can be used to extract the surface topographic profile of the sample after processing through a four-image phase shifting algorithm [12].

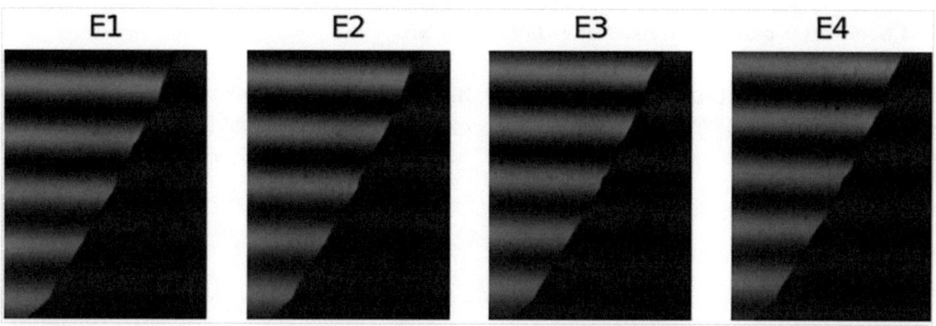

Figure 4.
Example of four DHs E_1, E_2, E_3, and E_4 with horizontal fringes obtained at phase 0, π/4, π/2, and 3π/4, respectively.

2. Phase shifting algorithms (PSA) and number of reference mirror positions

Where TAHs use a reference light to observe the encoded depth information of a sample, DHM uses PSA and phase mapping to extract the height information that is encoded in DHs. Notable examples of linear phase shift techniques are the standard N-step PSA [10], three-step PSA [16], and four-step PSA [17]. A continuous four-step algorithm with sinusoidal phase modulation [12] is also introduced, and a unique discrete four step PSA with sinusoidal phase modulation mapping is described. While higher frame count algorithms exist, there is a trade-off in acquisition and computation time [18]. To properly analyze PSAs, the time-domain low-coherence interference equation is modified to consider the phase shift, $\varphi(t)$, as follows:

$$I_s(x,y) = \bar{I}(x,y) + A(x,y)\,cos\,[\varphi(x,y) + \varphi(t)] \tag{5}$$

2.1 N-step reference mirror positions PSA

The N-Step PSA involves the acquisition of N DHs with N-1 different phase shifts between them. Typically, the phase shifts are evenly spaced and have values 0, 2π/N, 4π/N, ... , 2πm/N [10]. Each DH contains information about the phase difference between the two interfering beams at the time of acquisition. The interference pattern of the DH can be mathematically represented as follows:

$$I_{sn}(x,y) = \bar{I}(x,y) + A(x,y)\cos\left[\varphi(x,y) - \frac{2\pi m}{N}\right] \tag{6}$$

In Eq. (6), m is the index of phase shifts, m = 0,1, 2, ... , N-1. After obtaining the I_{sn} for each step (i.e., an image is taken at each step), the phase of the interferogram at every point on the surface under measurement can be reconstructed. The amplitude information from these N DHs is then combined in a mathematical algorithm to acquire the DH phase information. The phase map $\varphi(x,y)$ can be expressed as [10, 19]:

$$\varphi(x,y) = tan^{-1}\frac{\sum_{m=0}^{N-1} I_{sn}(x,y)\,sin\left(\frac{2\pi m}{N}\right)}{\sum_{m=0}^{N-1} I_{sn}(x,y)\,cos\left(\frac{2\pi m}{N}\right)} \tag{7}$$

2.2 Three-step reference mirror positions PSA

Three DHs acquired at three different, but evenly spaced-out, phase shifts are the minimum number of DHs needed to determine a sample's phase information [20]. The phase positions are 0, 2π/3, and 4π/3 and the relationship between $\varphi(x,y)$ and DH amplitude is:

$$\varphi(x,y) = tan^{-1}\frac{\sqrt{3}(I_{s1} - I_{s2})}{2I_{s0} - I_{s1} - I_{s2}} \tag{8}$$

I_{s0} through I_{s2} are the DHs collected at each phase position and represent the amplitude values of each DH that are dependent on the differential optical path lengths between reference and sample paths.

2.3 Four-step reference mirror positions PSA

To enhance the performance of phase calculations, a four-step PSA is often employed [17]. This algorithm offers improved accuracy and robustness compared to lower-step algorithms. The discrete phase shifts used are 0, $\pi/2$, π, and $3\pi/2$. The four acquired DHs are used to acquire the phase map as follows:

$$\varphi(x,y) = \tan^{-1}\frac{I_{s1} - I_{s3}}{I_{s0} - I_{s2}} \tag{9}$$

I_{s0} through I_{s3} represents the amplitude values of the frames collected at each phase position, like the three-step PSA.

2.4 Four-step reference mirror positions PSA with sinusoidal phase modulation

Adding more steps to the PSA has been proven to make DHM more accurate by improving its resilience to noise [21]. Five-, six- [22, 23], nine, and eleven-step algorithms were tested [21], but the impact on computation became intensive. An alternate approach to minimize noise is by allowing continuous motion of the PZA or modulating element [12]. The DH intensity is thus represented as follows:

$$I_{sn}(x,y) = \bar{I}(x,y) + A(x,y)\cos[\varphi(x,y) + \psi\sin(\omega t + \theta)] \tag{10}$$

The parameters ψ, ω, and θ are the modulating amplitude, angular frequency, and phase of the modulating signal. In summary, Dubois [12] can synchronize a continuously moving actuator with the camera capturing interferograms with modulation period, T, to create four images (E_1, E_2, E_3, and E_4) that follow [12]:

$$E_n(x,y) = \int_{(N-1)T/4}^{nT/4} I_{sn}(x,y,t)dt, n = 1,2,3,4 \tag{11}$$

Finding the solution to Eq. (11) involves expanding the integral into a summation of Fourier components by taking advantage of the Bessel function of the first kind. The derivation is discussed in detail by Dubois [12]. The modulation parameters ψ, ω and θ are selected *via* an optimization simulation to minimize noise and when noise is expected to be at a minimum the $\varphi(x,y)$ can be related to E_1, E_2, E_3, and E_4 as follows:

$$tan\,\varphi(x,y) = \frac{-E_1 + E_2 + E_3 - E_4}{-E_1 + E_2 - E_3 + E_4} \tag{12}$$

2.5 Four-step continuous position of reference mirrors and PSA with reduced phase step distance

The methods discussed above create phase maps that are heavily dependent on the phase noise of the system. High-phase noise can create abnormal features that do not coincide with the sample itself or result in phases that are inverted from what is expected. Phase noise can be minimized by reducing the amount of motion desired from the modulating device and by minimizing the deviation to the required phase periodicity.

In a system that uses a continuous phase modulator, unsynchronized or improperly calibrated modulation parameters add phase noise. Since the collection of images from a CMOS camera is inherently at discrete points in time, the phase between each image is set and based on the position of the phase modulator, in this case, a PZA. It is observed that Eq. (12) can be used to relate phase information to DH amplitudes when the phase difference between each DH is $\pi/4$. This is shown *via* simulation in Section 4. Acquiring DHs in discrete phase intervals has the advantage of simplifying the synchronization and collection of DHs. Another advantage is that $\pi/4$ phase change minimizes the needed PZA travel to collect 4 DHs over the $\pi/2$ phase change needed by Eq. (9) (i.e., 33.75 nm vs. 67.5 nm).

Using the results shown in Section 5 later, we conclude that four-step continuous position of reference mirror and PSA with reduced step distance is a practical solution to this DHM.

3. Phase unwrapping

The PSAs presented above allow for the extraction of phase for each pixel of the camera sensor. Constraints of the arctangent function force the extracted phase to be wrapped between a $-\pi$ to $+\pi$ distribution, leading to discontinuities in the observed phase. Phase unwrapping algorithms reconstruct the original phase signal from wrapped phase measurements, which are typically constrained to a 2π interval due to the periodic nature of phase. The goal of phase unwrapping is to reconstruct a smooth and continuous phase map that accurately represents the true underlying signal, even in the presence of noise or complex phase variations [24]. The best-known phase unwrapping method is a branch cut or path-following method while other methods are called minimum-norm methods [25]. The path-following method, referred to as Goldstein's algorithm, is selected for phase unwrapping for its speed and consensus in satisfactory unwrapping [26, 27].

A one-dimensional example of discontinuous data is extracted from the phase map generated using Eq. (12) and shown in **Figure 5** as unprocessed data. This phase map

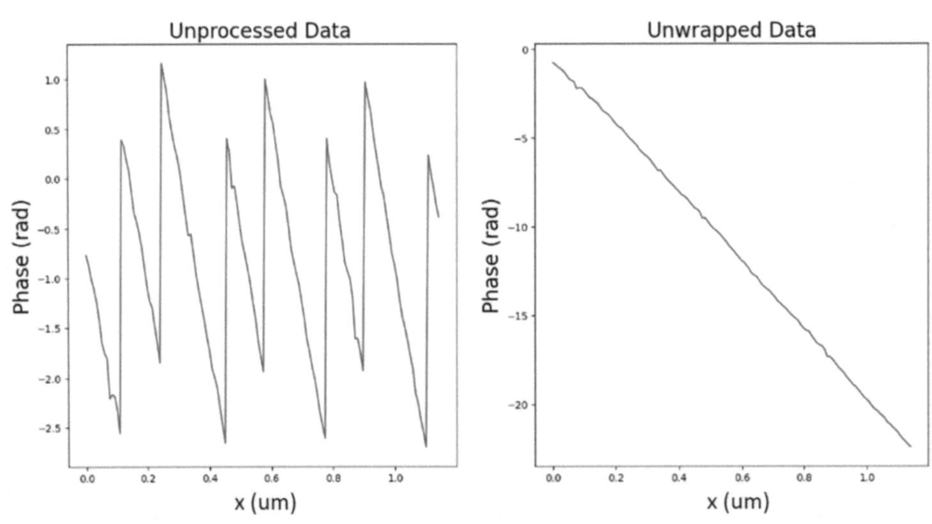

Figure 5.
Unprocessed and unwrapped one-dimensional phase map data as an example [28].

is constrained within the interval from -π to +π. The first step is to examine the locations where the phase gradient does not sum to zero in a closed loop [27]. If the difference between two successive phase measurements exceeds π, in magnitude, it is inferred that a phase jump or wrap has occurred. The exact number of full 2π wraps is identified by rounding the phase difference divided by 2π to the nearest integer [27].

Subsequently, a correction term, equivalent to the estimated number of wraps multiplied by 2π, is subtracted from the current phase value, effectively unwrapping the phase. This procedure is performed iteratively, progressing through the data sequence, usually in a 2×2-pixel grid. **Figure 5** unwrapped data is a smooth transition rather than a series of jumps, effectively transforming the phase data into a linear or near-linear form. The unwrapped phase values then provide an accurate representation of the continuous phase profile. Note that the exhibited unwrapped data has an inherent slope created by the unwrapping process.

4. Noise considerations of four-step PSA with π/4 phase difference

There are four significant sources of errors that have been identified to affect PSA-based DHM systems: phase error, motion error, intensity error, and intensity saturation error. These errors are discussed at length [29]. In this section, we provide a comprehensive review and discuss various behaviors exhibited by different error sources. Additionally, we discuss the potential solutions to mitigate these specific errors.

4.1 Phase errors

In an interferometric system, measurement accuracy is primarily influenced by phase errors, which significantly affect both precision and overall measurement quality. Phase errors cause discontinuities in the phase map that do not correspond to the phase differences resulting from the sample's physical characteristics [30]. As a result, algorithms that process data containing phase errors will have incorrect phase reconstructions. To reduce the impact of phase errors, a couple solutions are proposed. One approach is to increase the number of phase shift steps, enabling a more thorough analysis of the data [23, 31]. Alternatively, selecting a robust PSA, such as the four-step PSA [17] or the four-step PSA with sinusoidal phase modulation [12, 32, 33], can help minimize sensitivity to phase errors.

Simulations of the four-step PSA, in conjunction with Eq. (12), demonstrate that a reliable phase map can be generated when the phase signal-to-noise ratio (SNR) is set to 10 as shown in **Figure 6(a)** and **(b)**. In **Figure 6(a)**, when the phase SNR is reduced to 10, the sinusoidal fringe pattern along a row may exhibit a slight phase deviation. This deviation leads to phase oscillations in the resulting phase map, as shown in **Figure 6(b)**, where the phase values across the camera pixels (indicated by red x) deviate from the expected straight-line relationship (depicted in blue).

Phase-shifting techniques that involve a PZA can introduce phase noise due to the PZA's non-linear motion, inaccuracies in the step size calibration, or a misalignment between the camera and PZA frame capture timing. This phase noise can be mitigated by accounting for the PZA's characteristics, such as hysteresis and the properties of its control signal. As shown in **Figure 7**, the phase shift between DHs is influenced by the direction of the PZA's movement. To maintain consistent phase shifts during frame collection, it is crucial to capture DHs while the PZA moves in the same direction (either forward or backward). If the PZA changes direction, it is recommended to skip

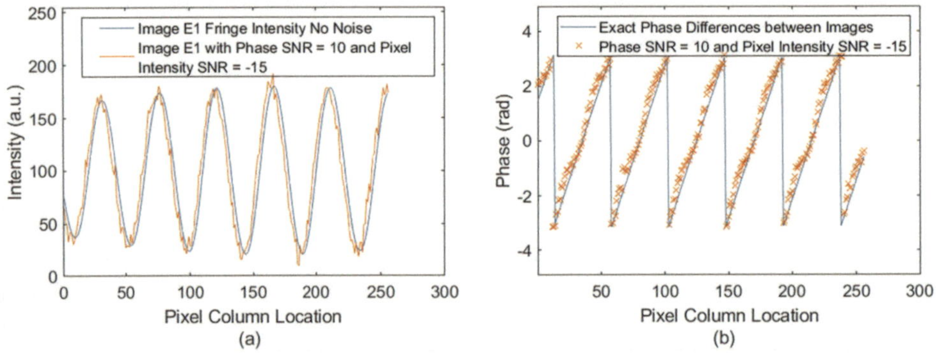

Figure 6.
(a) Phase of a single image row with and without noise. (b) Phase map created using four image PSA at π/4 phase shifts with and without noise [28].

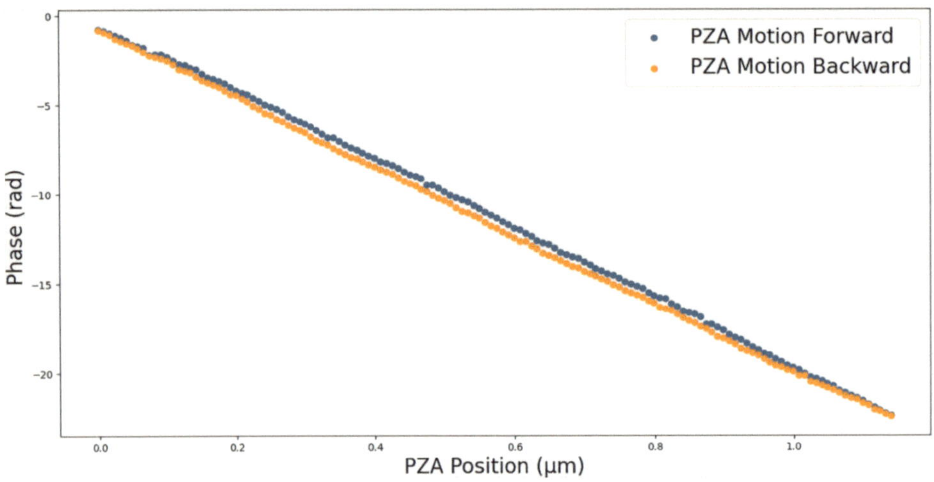

Figure 7.
PZA hysteresis as observed through fitted phases for DHs obtained through sweep of PZA position [28].

the first DH to avoid phase errors resulting from the different phase shifts in each direction.

An alternative approach to address the hysteresis of the PZA involves employing negative feedback to the PZA's input control voltage through a linear control system. By incorporating a highly accurate position sensor and voltage controller, the PZA can be precisely positioned before capturing a frame to be used as a DH as part of the four-step PSA. In practical applications, a simple proportional integral controller can effectively minimize phase error; however, it comes with the drawback of longer settling times, which in turn extends the overall acquisition time.

4.2 Motion, intensity, and intensity saturation errors

Motion, vibration of the sample, or vibrations of the DHM components introduce random errors during DH acquisition. These are distinct from the phase-shifting errors previously discussed. Mechanical vibrations can originate from sources such as a computer fan, cables connecting the computer to the optical system, or the system itself, potentially causing blurring or distortion of measured features. To mitigate random mechanical vibrations, the DHM system should be mounted on an optical table for vibration isolation and cables should be securely fixed to the table to prevent vibrational shorts to ground. Additionally, vibrations caused by air circulation can be minimized by enclosing the system in a single, compact unit.

Vibrational errors can result in both phase and intensity errors. Intensity errors occur due to fluctuations in the light intensity within the DHM, particularly when the reference beam intensity and sample beam intensity are not properly balanced. Such mismatches lead to inaccuracies in the measured phase difference. Factors contributing to intensity errors include variations in the light source's power and stability, imperfections in optical components, and differences in the reflectivity of the reference and sample surfaces. These issues ultimately impact the accuracy of phase difference measurements and the reconstruction of surface topography.

As illustrated in **Figure 8(a)**, when the reference arm is out of focus, the interference patterns on the step sample become increasingly noisy and unevenly distributed. This results in low-contrast fringe patterns even if the reference and sample optical path length are matched. High contrast fringes are observed in **Figure 8(b)** after the reference objective is placed into position to have the reference mirror in focus. High-contrast fringes have a large dynamic range and low-intensity noise.

Intensity saturation occurs when the camera's exposure setting is too high relative to the intensity of the reference and sample beams. Using saturated fringe patterns for PSA calculations introduces phase errors due to the truncation of fringe-pattern intensity, as shown in **Figure 9**. To prevent this issue, the camera exposure time should be carefully adjusted according to the surface reflectivity. By optimizing the exposure time, the camera can capture a balanced intensity range that avoids both underexposure and saturation, ensuring accurate measurements while maintaining

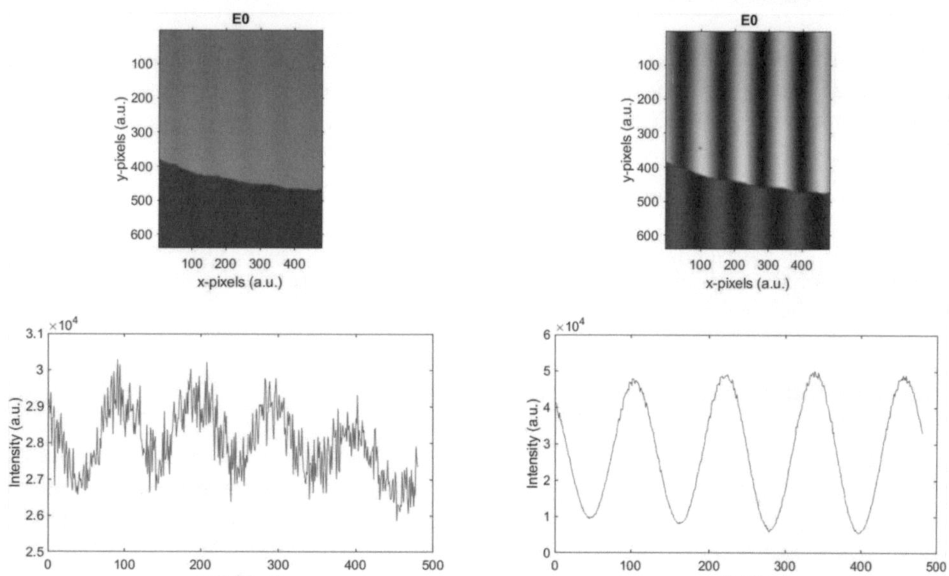

Figure 8.
(a) Intensity errors caused by out-of-focus reference position even though optical path length is matched. (b) In focus, the reference mirror produces DH with high-contrast fringes.

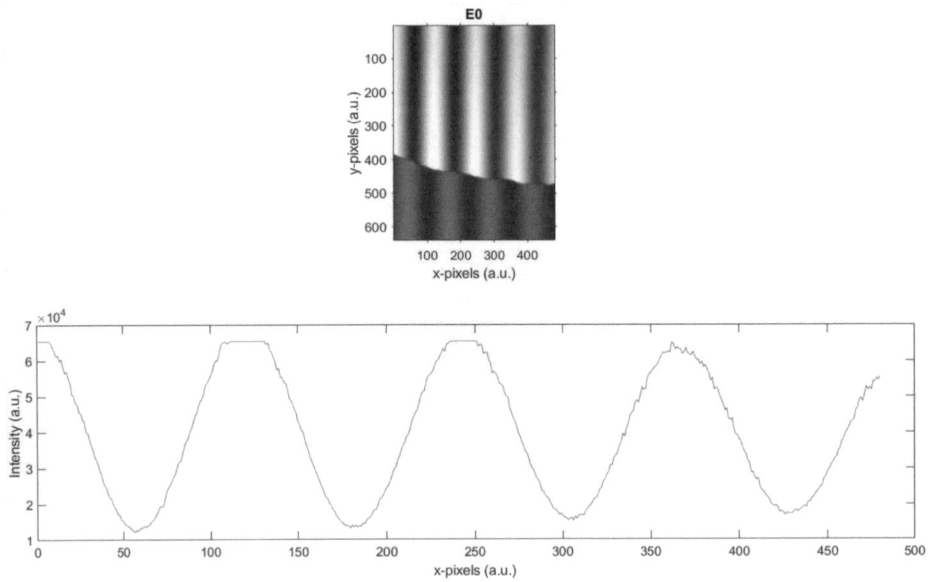

Figure 9.
Saturated DH pattern and the corresponding horizontal cross sectional intensity distribution showing truncation.

the quality of the fringe patterns. The camera gain should always be kept at zero. Any nonzero gain will magnify intensity noise.

5. Application of DHM for non-contact surface profiling

The constructed DHM, as 3D schematically illustrated in **Figure 2**, has been programmed for the effective PSA and tested for a non-contact surface profiling application. The system employs a four-step PSA, maintaining a controlled phase difference of $\pi/4$ between consecutive DHs. Optimal performance is achieved by determining the actual interferometric characteristics of the light source, as illustrated in **Figure 3(c)**, to establish the PZA resolution. Once the minimum PZA motion is identified, the camera is synchronized to capture four DHs, as shown in **Figure 4**, enabling the characterization of a sample. The results are then compared to the DektakXT surface profiler, a known standard, for validation.

5.1 PZA resolution characterization

The minimum phase shift achievable by the PZA was determined by mapping the applied PZA voltage to the coherence bandwidth scan of the light source. A series of voltage changes in 1 mV increments was applied, totaling 42,000 scan points. The wavelength parameters λ_0 540 nm and $\Delta\lambda$ 40 nm were obtained using a USB2000 spectrometer from Ocean Optics.

From this analysis, the PZA's minimum resolution was calculated to be 11.81 nm per 1 mV, corresponding to a phase shift of approximately 0.275 radians or 15.8° per 1 mV. To achieve the desired $\pi/4$ phase difference between consecutive DHs, a differential voltage change of approximately 2.86 mV is determined to be optimal.

5.2 Profiling of step samples

Five-step samples of varying heights were fabricated to evaluate the performance of the DHM surface profiler. Each step sample consisted of a chromium layer deposited on fused silica. The chromium was evaporated onto the fused silica, followed by spin-coating a layer of photoresist onto the chromium. Photolithography and chromium etching were then performed to create step windows. The different step heights were achieved by controlling the etch times.

The step height ranges of the samples were first measured using the DektakXT surface profiler. For each sample, measurements were taken from 12 different step regions, and the averages of these measurements were used for comparison. The corresponding average step heights measured by the DHM were also obtained and shown in **Figure 10** with the error bars. The relationship between unwrapped phase map and height is provided by Eq. (4).

A strong positive correlation coefficient of 0.9984 was observed between the measurements of DHM and DektakXT. The maximum measured standard deviation with the DektakXT was 3 nm, observed in the 94 nm step height sample, and this variability was attributed to the actual thickness variation across different locations of the sample. The average surface height measurements obtained with the DHM were within 3 nm of the DektakXT measurements. This discrepancy is attributed to system noise, non-optimized PZA calibration, and position-dependent thickness variations. The present height resolution of the DHM, considering noise, is 3.3 nm.

Figure 11 presents the 3D reconstruction of the ~125 nm step sample, created using the DHs E_1 through E_4 from **Figure 4**. This step profile was determined by calculating the height difference between the higher and lower sections of each row and then averaging these values. This process was repeated for all five-step samples to ensure a comprehensive measurement. **Figure 11** also shows the slope created by unwrapping the phase map removed by trend removal functions.

The ~125 nm step sample was further analyzed using various existing discrete step PSAs. **Table 1** provides detailed height measurements obtained with each algorithm. **Figure 12** illustrates the 3D reconstructions of the ~125 nm step sample generated

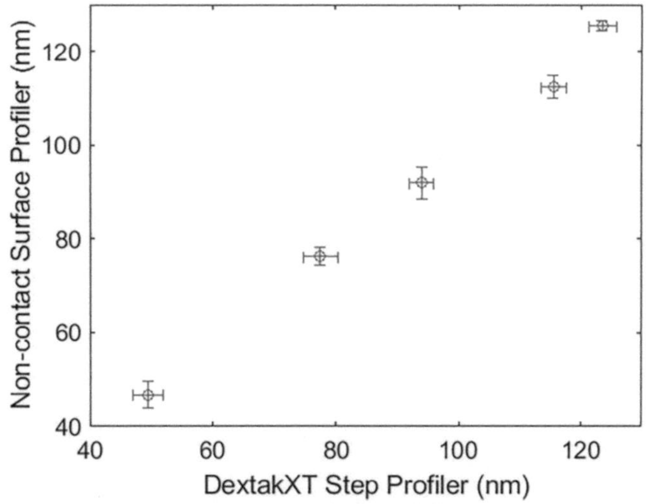

Figure 10.
Surface step height measurement comparison between our DHM and DektakXT line scan step profiler [28].

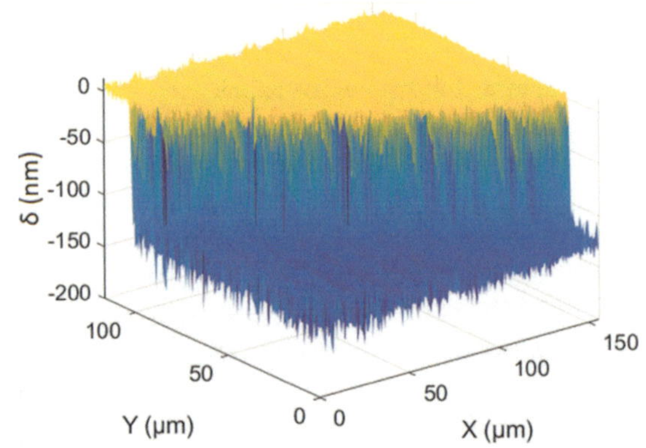

Figure 11.
The 3D reconstruction of a step surface pattern generated using E_1, E_2, E_3, *and* E_4 *from* ***Figure 4****. The step has an average profiled height of 123.7 nm, width of 117.12 μm, and length of 156.16 μm. The slightly wavy surface is due to the measurement noise of the system.*

	Height evaluation	
Phase shifting algorithm applied	**Average step height**	**Error**
Three-step	99.0 nm	4.7 nm
Four-step	115.0 nm	9.0 nm
Four-step continuous PSA with reduced step distance (our method)	125.6 nm	3.3 nm

Table 1.
Height of 125 nm step sample using different PSA [26].

using the different PSAs. The four-step PSAs exhibit noticeable phase-related noise on the reconstructed surface (wavy surface). These phase-related errors, including oscillations along the x-axis and incorrect step height correlations, are attributed to improper phase mapping caused by the limitations of the arctangent function. The slanted surface (not yet leveled by trend removal functions) is due to a tilted reference mirror for the creation of the needed interference fringes.

The impact of fringe orientation on the reconstructed image and the measured thickness was also analyzed showing fringe orientation independence.

5.3 Noise considerations and axial and lateral resolution of the surface profiling system

To evaluate phase noise, several images were collected without PZA motion, and the phases of the interference patterns were fit to a sinusoidal function. The standard deviation of the phase was calculated as 0.040 rad, corresponding to a system noise error of approximately 1.72 nm using Eq. (4). Subsequently, the PZA was actuated through a series of $\pi/4$ intervals, with variations between intervals measured at 0.037 rad, equating to a PZA positional noise of 1.59 nm again using Eq. (4). Combining these results, the total axial resolution was determined to be approximately 3.31 nm.

The lateral resolution of the DHM system was assessed using a 1951 USAF Resolution Target from Thorlabs, placed under the microscope as a standardized target. The smallest group of lines was identified, and the horizontal and vertical bars in Element

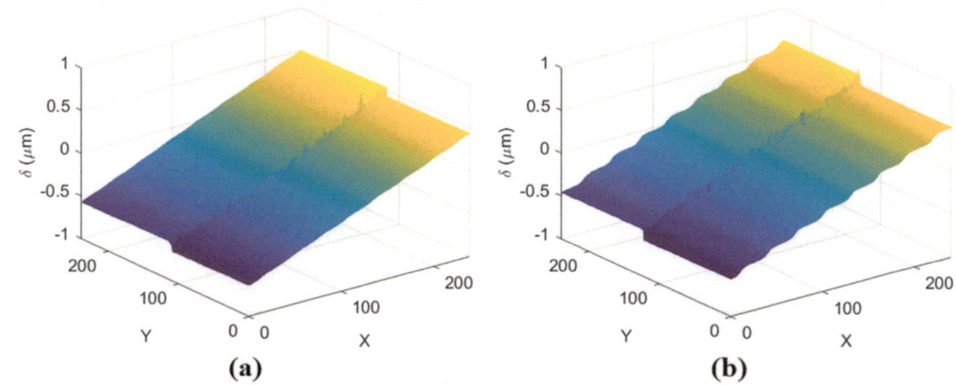

Figure 12.
3D reconstruction of the 125 nm step surface using (a) three-step and (b) four-step PSA. Unwanted oscillation is observed along the x-axis for the four-step method, and measured height does not correlate with the sample's actual thickness [26].

6, Group 6, were clearly resolved. Analyzing the number of pixels between the lines in this element, the lateral resolution was calculated to be 488 nm.

5.4 Leveling and complex samples

The phase unwrapping process in the DHM system introduces a consistent trend across the 3D reconstructed images, extending the δ – axis to the microscale. This trend is evident in **Figure 12** and results from the tilting of the reference mirror used to generate interference fringes. While mirror tilting is essential for creating the

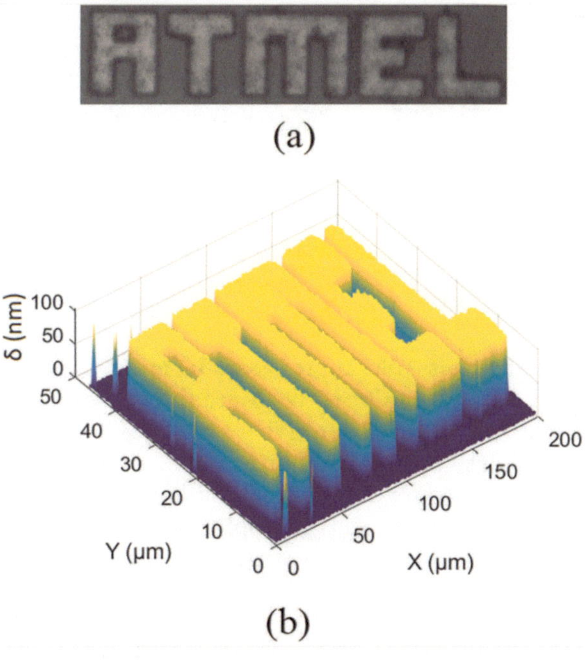

Figure 13.
(a) Conventional microscope image and (b) 3D DHM image of a memory chip showing ATMEL word.

fringes, the resulting tilt in the reconstructed 3D surface image should be removed to facilitate user-friendly visualization of the sample surface.

This process, commonly referred to as leveling, is standard in most profiling systems. MATLAB's detrend function is employed to identify and remove the linear trend from the data. An example of this linear trend is shown in **Figures 5** and **12** unwrapped data. Once the trend is removed, the true scale of the 3D reconstructed images becomes more apparent. This is demonstrated in **Figure 13**, where (a) shows the conventional 2D microscope image and (b) the 3D DHM image of the word "ATMEL" on the surface of a computer memory chip with a height of approximately 100 nm.

6. Conclusion

DHM is a powerful technique that combines advanced imaging and quantitative analysis capabilities, offering unique benefits for non-contact measurement. The discrete-step phase-shifting method enables high-resolution depth measurements by precisely capturing the phase differences across pixels. This approach ensures accurate reconstruction of the surface topography and material properties of reflective samples, even at the microscopic level. DHM should gain popularity in diverse fields such as materials science, semiconductor inspection, biomedical research, and nanotechnology. Its ability to measure surface profiles, detect defects, and analyze thin films or biological structures without physical interaction makes it invaluable for sensitive or delicate samples. The non-contact nature would eliminate the risk of sample damage, contamination, or deformation, which is especially important for applications requiring precise and repeatable measurements.

The integration of discrete step phase-shifting in reflective digital holography microscopy significantly enhances its measurement precision and reliability. The method's non-invasive nature, coupled with its versatility, positions DHM as a critical tool for advancing research and industrial applications where high-resolution, non-contact surface analysis is essential. As technology progresses, DHM is poised to further transform how we approach complex measurement challenges across multiple disciplines due to the inherent ability of reducing noise with the addition of more phase-shifted digital holograms.

7. Summary

DHM systems offer a robust solution for areal 3D microscopic surface measurements, making them an excellent alternative to traditional surface profilers.
The optical setup features a single aperture that facilitates both illumination and image capture, a critical design choice for handling the spatial constraints of nanoscale objects. By reconstructing topographic surface profiles without physical contact, DHM eliminates the risk of damage to the sample under test and eliminates maintenance tasks like stylus replacements often associated with contact-based scan profilers.

A four-step PSA with discrete image selection to reconstruct 3D surface profiles of objects with varying thicknesses was demonstrated. Persistent oscillations in reconstructed 3D images, observed when using sinusoidal phase modulation, were minimized by limiting the phase difference between DHs to $\pi/4$. These oscillations were also observed in three-step and four-step discrete PSAs, and we compared the accuracy of height measurements for each PSA and their respective phase differences. Additionally, the total PZA motion required was reduced by employing a smaller phase difference.

The surface height resolution of the DHM system is approximately 3.31 nm, while the lateral resolution, determined by the magnification of the objective lens, is 488 nm. The current FOV, defined by the magnification lens and camera format, measures 312 μm × 234 μm. While lateral resolution can be enhanced using a higher magnification lens, this improvement comes at the cost of a smaller FOV and reduced fringe pattern contrast.

The non-contact DHM generates a 3D surface profile within 52 ms, with processing time dependent on the computer processor and PZA speed. Performance could be further improved with faster processors or PZA units. DHM has been implemented in a setting where the sample can change and allows for the digital storage of DHs instead of physical holographic film plates. The recorded DHs can facilitate online sharing with remote interested users.

Acknowledgements

The research as described in this chapter is supported by the National Science Foundation; Award No. CBET-2209271.

Author details

Diego Palacios and Michael R. Wang*
Department of Electrical and Computer Engineering, University of Miami, Coral Gables, FL, USA

*Address all correspondence to: mwang@miami.edu

References

[1] Kim MK. Principles and techniques of digital holographic microscopy. SPIE Reviews. 2010;**1**(1):18005

[2] Marquet P, Depeursinge C, Magistretti PJ. Exploring neural cell dynamics with digital holographic microscopy. Annual Review of Biomedical Engineering. 2013; **15**:407-431

[3] Pavillon N, Kühn J, Moratal C, Jourdain P, Depeursinge C, Magistretti PJ, et al. Early cell death detection with digital holographic microscopy. PLoS One. 2012;7(1):1-9

[4] Rappaz B, Breton B, Shaffer E, Turcatti G. Digital holographic microscopy: A quantitative label-free microscopy technique for phenotypic screening. Combinatorial Chemistry & High Throughput Screening. 2014;**17**(1): 80-88

[5] Gao P, Yuan C. Resolution enhancement of digital holographic microscopy via synthetic aperture: A review. Light. Advanced Manufacturing. 2022;**3**(1):105

[6] Vabre L, Dubois A, Boccara AC. Thermal-light full-field optical coherence tomography. Optics Letters. 2002;**27**(7):530-532

[7] Dubois A, Vabre L, Boccara AC, Beaurepaire E. High-resolution full-field optical coherence tomography with a Linnik microscope. Applied Optics. 2002;**41**(4):805-812

[8] Ogien J, Dubois A. A compact high-speed full-field optical coherence microscope for high-resolution in vivo skin imaging. Journal of Biophotonics. 2019;**12**(2):e201800208

[9] de Groot P. Principles of interference microscopy for the measurement of surface topography. Advances in Optics and Photonics. 2015;7(1):1-65

[10] Srinivasan V, Liu HC, Halioua M. Automated phase-measuring profilometry of 3-D diffuse objects. Applied Optics. 1984;**23**(18):3105-3108

[11] Drexler W, Fujimoto JG. In: Drexler W, Fujimoto JG, editors. Optical Coherence Tomography: Technology and Applications. 2nd ed. New York: Springer Cham; 2015. pp. 78-81

[12] Dubois A. Phase-map measurements by interferometry with sinusoidal phase modulation and four integrating buckets. Journal of the Optical Society of America. A, Optics, Image Science, and Vision. 2001;**18**(8):1972-1979

[13] Illarramendi MA, Zubia J, Arrue J, Ayesta I. Adaption of the Michelson interferometer for a better understanding of the temporal coherence in lasers. In: ETOP 2017 Proceedings. Hangzhou China: Optica Publishing Group; 2017. p. 1045249

[14] Judge TR, Bryanston-cross P. A review of phase unwrapping techniques in fringe analysis. Optics and Lasers in Engineering. 1994;**21**:199-239

[15] Trouvé E, Caramma M, Maître H. Fringe detection in noisy complex interferograms. Applied Optics. 1996;**35** (20):3799-3806

[16] Zhang S. Composite phase-shifting algorithm for absolute phase measurement. Optics and Lasers in Engineering. 2012;**50**(11):1538-1541

[17] Omidi P, Najiminaini M, Diop M, Carson JJL. Single-shot 4-step phase-shifting multispectral fringe projection

profilometry. Optics Express. 2021;**29**(18):27975-27988

[18] Schmit J, Creath K. Extended averaging technique for derivation of error-compensating algorithms in phase-shifting interferometry. Applied Optics. 1995;**34**(19):3610-3619

[19] Bruning JH, Herriott DR, Gallagher JE, Rosenfeld DP, White AD, Brangaccio DJ. Digital Wavefront measuring interferometer for testing optical surfaces and lenses. Applied Optics. 1974;**13**(11):2693-2703

[20] Hibino K, Oreb BF, Farrant DI, Larkin KG. Phase shifting for nonsinusoidal waveforms with phase-shift errors. Journal of the Optical Society of America. A. 1995; **12**(4):761-768

[21] Shen MH, Hwang CH, Wang WC. Using higher steps phase-shifting algorithms and linear least-squares fitting in white-light scanning interferometry. Optics and Lasers in Engineering. 2015;**66**:165-173

[22] Su X, Chen W. Reliability-guided phase unwrapping algorithm: A review. Optics and Lasers in Engineering. 2004; **42**:245-261

[23] Zappa E, Busca G. Comparison of eight unwrapping algorithms applied to Fourier-transform Profilometry. Optics and Lasers in Engineering. 2008;**46**: 106-116

[24] Li C, Zhu D. A residue-pairing algorithm for Insar phase unwrapping. Progress in Electromagnetics Research-Pier. 2009;**95**:341-354

[25] Ghiglia DC, Pritt MD. Two-Dimensional Phase Unwrapping: Theory, Algorithms, and Software; 1998

[26] Zhai B. Optical Interferometric Microscopy Using Controlled Phase Shifting Algorithms [Master's Thesis]. Coral Gables, FL: University of Miami; 2023

[27] Zuo C, Feng S, Huang L, Tao T, Yin W, Chen Q. Phase shifting algorithms for fringe projection profilometry: A review. Optics and Lasers in Engineering. 2018;**109**:23-59

[28] Palacios D, Zhai B, Wang MR. Noncontact surface profiling using optical interferometric microscopy. Optical Engineering. 2024;**63**(2):23104

[29] Zhao B, Surrel Y. Phase shifting: Six-sample self-calibrating algorithm insensitive to the second harmonic in the fringe signal. Optical Engineering. 1995; **34**(9):2821-2822

[30] Chen Y, Kemao Q. Phase-shifting algorithms with known and unknown phase shifts: Comparison and hybrid. Optics Express. 2022;**30**(5):8275-8302

[31] Dubois A, Boccara AC, Lebec M. Real-time reflectivity and topography imagery of depth-resolved microscopic surfaces. Optics Letters. 1999;**24**(5):309-311

[32] Dubois A, Grieve K, Moneron G, Lecaque R, Vabre L, Boccara C. Ultrahigh-resolution full-field optical coherence tomography. Applied Optics. 2004;**43**(14):2874-2883

[33] Van EAG, Diddams SA, Clement TS. Dispersion measurements of water with white-light interferometry. Applied Optics. 1998;**37**(24):5679-5686